Potash Use and Dynamics in Agriculture

Abdul Wakeel • Muhammad Ishfaq

Potash Use and Dynamics in Agriculture

Springer

Abdul Wakeel
Institute of Soil and Environmental Sciences
University of Agriculture
Faislabad, Pakistan

Muhammad Ishfaq
College of Resources and Environmental Sciences
China Agricultural University
Beijing, China

ISBN 978-981-16-6882-1 ISBN 978-981-16-6883-8 (eBook)
https://doi.org/10.1007/978-981-16-6883-8

This Springer imprint is published by the registered company Springer Nature Singapore Pte Ltd.
The registered company address is: 152 Beach Road, #21-01/04 Gateway East, Singapore 189721, Singapore

This book is dedicated to Professor Konrad Mengel (1929–2012) from Justus Liebig University, Giessen, Germany, for his legendary work on potassium nutrition of plants.

Foreword

Pakistan is an agricultural country, and a major share of its economy is dependent on agriculture. However due to mismanagement and lack of systematic agricultural research, agricultural production is declining very rapidly. There is a need to promote problem-based research in Pakistan for which there is a dire need of capacity building based on precise subject-related and site-specific data. The book in hand is a specialized book with comprehensive knowledge of potassium nutrition starting from its behavior in soil to its impact on produce quality leading to human health. It is obviously a concise and comprehensive book ever before which has concluded the scope of potassium nutrition in a beautiful way.

The first chapter of the book provides an introduction to potassium with recent updates in potassium research developments, followed by its behavior and dynamics in soil in the next chapter. The important thing is that it has focused the Pakistani soils in global scenario. Chapter 3 throws light on some scholarly aspects of the physiological role of potassium in plants, followed by Chap. 4 focusing on the phytoavailability of potassium for plants which is a complete guidance for modern research aspects related to potassium nutrition. The fifth chapter is about the determination of potassium in soil, which has a key role to understand the potassium dynamics in soil and phytoavailability of potassium to plants. The sixth chapter explains the role of potassium to maintain and enhance the quality of cereals, sugar crops, vegetables, and fruits considering nutrition as well as aesthetic aspects of agricultural commodities, while seventh chapter grasps the biofortification aspect of potassium nutrition in staple crops in the light of contemporary studies. Chapter 8 usefully and critically covers the potassium research in Pakistan, considering reasons of low use of potassium fertilizers and the strategies to improve its judicious use. The key chapter is the last chapter which concludes with the future research options for Pakistani agriculture, directing researchers not only working in Pakistan but also at the global level. This book will open new avenues in agricultural research in Pakistan.

I believe that this book is the first-ever comprehensive book on potassium nutrition equally beneficial for undergraduate and research students. Furthermore, for extension workers in the public and private sector, this manuscript will be a big asset in the upcoming decade. Indeed, I appreciate the efforts of the authors who have spent their precious time compiling this valuable information.

Sultan Qaboos University
Muscat, Oman

Muhammad Farooq

Preface

Potassium (K), an essential plant nutrient, has long been overlooked in agriculture in many developing countries. In most of the agroecosystems of these countries, potash balance is negative because its application seldom matches with removal. The current imbalanced fertilization practices without K fertilizer application led to its mining, which results in certain negative impacts, including low nutrients use efficiency such as nitrogen, decreasing farmers' income ultimately. At the same time our agro-technicians lack enough skills and resources to promote the right source of fertilizer at the right rate, time, and place to facilitate profitable farming. Therefore, farmers need to update their farming practices to improve their crop yield and quality under expanding unfavorable climatic conditions to feed huge population of the developing countries.

Food security and agriculture sustainability depend on how small farmers, who are typically resource-limited and isolated from advanced technology, manage their farming practices. The evidence in support of K deficiency is rising in these countries. Combined approaches are needed to pursue balanced fertilization program in developing countries. Building trust, participatory innovation, developing human capacity, and creating awareness among local farming communities are critical to uplift the agricultural economy.

There is a great potential to increase crop yield per unit land area in most of the developing countries by application of K fertilizers. However, crop response to K fertilization cannot always be linear, and there is a need to attain the site-specific K benchmark based on distinct soil type and minerology. It is important to effectively promote the significance of balanced fertilization at grassroots level for the benefit of the farming community. In this book, as far as possible, the consequences of K nutrition under the current agriculture scenario are highlighted stating K nutrition for soil, plants, and humans. It is emphasized that to ensure economic agriculture sustainability, balanced and site-specific potash recommendations are needed in developing countries such as Pakistan. This book contains useful information equally demanded by graduate students, progressive farmers, extension workers, and policy makers. I am very thankful to Dr. Hafeez ur Rehman and Dr. Ahmad Nawaz for their contributory support to complete the draft of this book.

Faisalabad, Pakistan   Abdul Wakeel

About the Book

This book covers the use and dynamics of potassium fertilizers in agriculture. It explores potassium dynamics in soil, phytoavailability, uptake and translocation in crop plants, impact of potassium fertilizers on the quality of agricultural produce.

Potassium is an essential plant nutrient that has long been overlooked in agriculture of many developing countries. In most of the agro-ecosystems of such countries, potassium balance is negative because its application seldom matches with crop removal. Agro-technicians lack enough skills and resources to promote the right source of fertilizer at the right rate, time, and place to facilitate profitable farming. There is a need for farmers to update their farming practices so as to improve the crop yield and quality under unfavorable climatic conditions. Correct application of potassium fertilizers is directly linked with increased crop yield per unit land area in most of the developing countries. Therefore, this book fills the gap in the information and provides the readers with the latest updates on the use of potassium fertilizers.

This book contains the latest information relevant for graduate students, progressive farmers, extension worker, early career researchers, and policymakers.

Contents

About the Authors

Abdul Wakeel is an Associate Professor (Tenured) at University of Agriculture Faisalabad. He received PhD degree in agriculture from Justus Liebig University Giessen, Germany, and worked there as a postdoctoral researcher afterwards with Prof. Sven Schubert. He also worked with Prof. Konrad Mengel and shared his legendary experiences of potassium nutrition. Dr. Wakeel coordinated the research and extension activities of International Potash Institute Switzerland in Pakistan for several years and has extensive experience of potash fertilization in Pakistan and south Asia. Dr Wakeel is working on sustainable nutrient management under changing climate. Currently, he is the chairman of Pakistan Agricultural Scientists Forum and managing editor of *Pakistan Journal of Agricultural Sciences*.

Muhammad Ishfaq is a researcher at the College of Resources and Environmental Sciences, China Agricultural University, Beijing, China. His major research interests are plant–soil interactions, plant physiology, root biology, soil health, and molecular plant nutrition. He graduated in soil science from the University of Agriculture Faisalabad, Pakistan, and worked as a research associate to promote precise and balanced use of fertilizers in Pakistan. Ishfaq is actively engaged in different research and developmental activities and published several peer-reviewed SCI papers in reputed journals. He was nominated for International Plant Nutrition Institute (IPNI) Scholar Award 2018 in recognition of an outstanding scholastic record and appreciation of contribution to the agricultural sciences.

Introduction

1

Since the 1960s, the world population has increased from 3 to 7.8 billion, and this trend of exponential expansion will persist in the coming decades. Accordingly, a substantial increase in agriculture production is needed to feed future generations on sustainable basis. Additionally, recent climate models predict that incidences and duration of heat and drought stress are expanding in several regions of our planet which negatively affecting crop production, and ultimately food security. In consequence, at present leading challenges are to improve agriculture productivity in extra resource-efficient ways and to stabilize crop yield under various environmental stress conditions. In this context, potassium (K) nutrition has revealed promising potential to promote numerous biological processes crucial for plant growth, yield, quality traits, as well as stress tolerance in the field.

Potassium is an indispensable element for crop and human health being involved in various critical metabolic and physiological functions. It is soft, silver-white metal, light in its pure form, seventh most abundant element and contributes 2.1–2.3% share of earth crust. Its atomic number and atomic weight are 19 and 39, respectively, and in the periodic table, it is included "alkali metals" group. In plants, it mediates numerous fundamental functions and needed in large amounts to fulfill their growth requirements, for instance, it involves in photosynthesis, activation of more than 60 plant enzymes, stomatal regulation, protein synthesis, osmotic adjustment, and phloem transport (Marschner 2012; Wang and Wu 2017). It plays a fundamental role in the ability of plants to tolerate externally induced stresses such as drought, frost, highlight levels, and incidence of pests and diseases (Wang et al. 2013). Therefore, growth and productivity of agriculture are highly dependent on the availability of K.

The application of K containing fertilizers improves the quality of agriculture produce. In sugar beet, the use of K fertilization can improve the yield, growth, and quality traits (Etemadi 2000; Ouda 2002; Mubarak et al. 2016). Furthermore, K fertilization in sugar beet can improve N, P, K, total soluble sugar, purity percentage, and sucrose contents (Amer et al. 2004). Additionally, K also plays a very important

A. Wakeel, M. Ishfaq, *Potash Use and Dynamics in Agriculture*,
https://doi.org/10.1007/978-981-16-6883-8_1

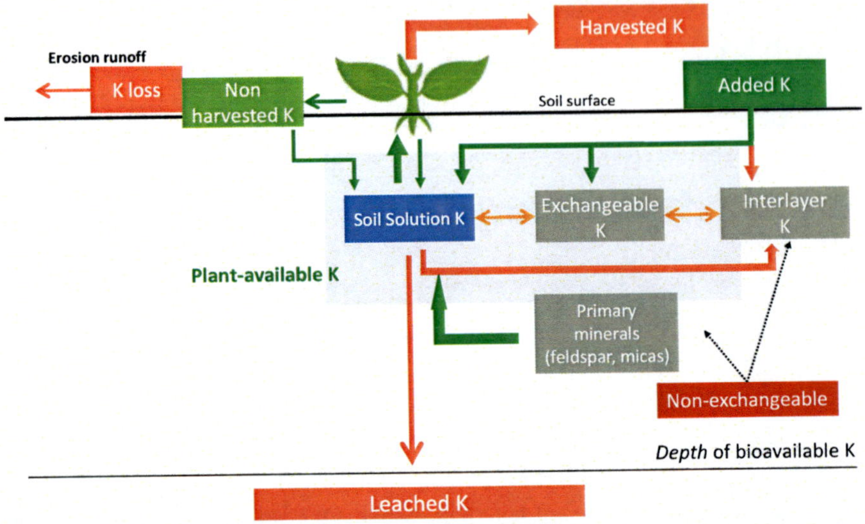

Fig. 1.1 The potassium cycle in the soil–plant system as drafted at the workshop: "Frontiers in Potassium Science: Developing a Roadmap to Advance the Science of Potassium Soil Fertility Evaluation," 26–30 January, 2015 in Kailua Kona, Hawaii in conjunction with the International Symposium on Soil and Plant Analysis (http://www.isspa2015.com/)

role under adverse soil conditions. Additional potash application can improve seed-cotton yield of early maturing Bt cotton under water deficiency (Shahzad et al. 2019). Furthermore, it can reduce the deleterious impact of salt stress on plant and ultimately improve crop productivity (Chen et al. 2007; Wakeel 2013; Gul et al. 2019). Therefore, it is imperative to optimize potash recommendations under adverse soil conditions for profitable crop production.

Potash availability and spatial distribution in soils relies on a wide range of intrinsic and extrinsic factors. Understanding how K availability varies in space and time is decisive for preventing potential deficiencies. To describe the complex K dynamics in the soil, K in the soil is usually categorized into four groups in terms of its availability for plants: (a) soluble K present in soil solution, (b) exchangeable K, (c) non-exchangeable, and (d) structural K (Rehm and Schmitt 2002). The readily available pools of K for plants are considered, K dissolved in soil solution and exchangeable fraction (held on the sites of clay minerals or soil colloid), and its uptake may vary from 0.10 cmol kg^{-1} to 0.40 cmol kg^{-1} according to crop type and growth environment (Dobermann et al. 2004; Dunn and Stevens 2005; Koné et al. 2011). The measurement of readily available pool of K is usually considered valuable to predict soil K status in relation to plant supply. However, the major share of the soil K is in the structural form, largely contained of K-bearing primary minerals, for instance, biotite, muscovite, and feldspars. Recent deliberations on assessment of plant available soil K content have revealed that non-exchangeable K also contributes significantly to plant growth. Keeping in view these advancements, K cycle was proposed to be modified in 2015 (Fig. 1.1) and then in 2017 it was endorsed during a global meeting on potassium in Rome. Therefore, assessment methods need to be revised to include the non-exchangeable fraction of K to be assessed. Some methods have already been devised for such assessment; however, an intensive and site-specific research has been suggested during various sessions.

The inputs of K in soil can be chemical fertilizers, manure, crop residues or irrigation water, especially canal water while output is crop removal or K losses through leaching, in sandy soils, and runoff. The input should be balance with output to attain optimum quantity and quality of food production on a sustainable basis. Although soil parental rock minerals are also the source of K to some extent, but their K releasing capacity depends on the degree of weathering and mineralogical composition of soil. Soils characteristically low in K are commonly sandy, water-logged, acidic, or saline. Furthermore, soil mining with continuously intensive cropping with lower or unbalanced fertilization can diminish the indigenous soil K contents, for instance, 3/4 of the paddy soils of China (especially Southern China) and 2/3 of the wheat belt of Southern Australia are declared as K deficient (Mengel and Kirkby 2001; Römheld and Kirkby 2010).

Apart from plants, K also mediates several physiochemical processes in humans, and plants are the basic source of K nutrition for both human and animals. Therefore, the basis of human and animal health-relies on the supply of adequate amount of nutrients via agricultural commodities and soil is the initial point. In human, K performs many functions such as it participates in ordinary nerve transmission, osmotic pressure regulation, heart function, muscle contraction, kidney, and bone health as well as it encompasses in secretion and action of hormones, that control orderly blood pressure, acid–base homeostasis, mobility of gastrointestinal and glucose, and insulin metabolism (Weaver 2013; Calabrese et al. 2019). Low K level (hypokalemia) can cause weakness, muscles pain, lack of energy, stomach disturbance like symptoms as well as it can cause stroke, arthritis, and inflammatory bowel disease relevant disorders. The use of high amount of Na in our diet as compared to K can raise blood pressure and resultantly lead to cardiovascular diseases. Therefore, it is prerequisite to improve K intake in our diet as K-rich fruits and vegetables are the safest and preferred pathway to improve K intake.

Many years back it is proposed that plants grown on soils rich with exchangeable K appears no symptoms of K deficiency for a certain time period of plant growth (Mengel and Kirkby 1982). However, the soils not having sufficient exchangeable K sites might be susceptible to K deficiency, thus affecting the final crop yield. Indeed, K deficiency lead to low starch level, poor formation of reproductive structures, and hampered grain filling. It is revealed that adequate K nutrition can enhance protein content in wheat and barley (Holland et al. 2019). Furthermore, the improvement in plant height and biomass of wheat and maize are associated with K application. Additionally, the optimum supply of K can improve the efficiency of other mineral nutrients, for instance, nitrate (NO_3^-). In brief, K is an imperative yield limiting macronutrient, and potash fertilization is obligatory, especially in intensive agricultural system to optimize crop yield (Mussgnug et al. 2006).

Pakistani soils are largely alluvial in nature and developed from mica minerals which are known to be high in K content. Although, the total quantity of K is ~2.7%, the majority of this is not readily available to plants, owing to its strong binding with the primary and secondary clay minerals. The K fractions exist in soil solution, and at exchange sites are readily available for plants and to some extent non-exchangeable K can also be extracted by plant roots. Sometimes plants do not respond to the recommended amounts of potash application (Mengel and Kirkby 2001), due to close association of K with soil minerology, and applied K can be engaged in the interlayer empty spaces of the exhausted mineral. For instance, soils dominant in clay minerals, namely vermiculite and illite restrict the K uptake by plants due to their high cation exchange capacity (Mengel and Kirkby 2001) and K-fixing capacity. In contrast, a soil having a major share of trioctahedral micas, such as phlogopite and biotite, is testified to be more readily K releasing through weathering process.

In Pakistan, the use of K containing fertilizers is still under debate, as various crops show response to applied potash in the terms of growth and yield. Most of the potash recommendations are based on soil exchangeable K contents, while disregarding the spatial soil mineralogy and the actual K availability of distinct soil, thus sometimes blanket fertilizer recommendations lead to uncertain crop response to applied K. However, evidences in the support of crop response to applied potash are expanding in Pakistan, as nearly 40% soils of Pakistan are predicted to be K deficient (Akhtar et al. 2003). Therefore, K deficiency symptoms have been experienced in various crops, and many farmers are showing interest in K fertilization due to considerable yield difference. For instance, K application in sandy loam and sandy clay loam soils improved maize growth, whereas on clay loam soils, potash fertilization was not effective for improvement in plant growth, owing to K fixation or adequate indigenous soil K content (Wakeel et al. 2002). However, still much remain open to elucidate such as geographical distribution of K in line with soil minerology and soil–plant spatial response to attain a site-specific potash benchmark in arable farming systems of Pakistan.

In brief, although abundant in soils, K availability to plants may be suppressed depending on soil properties and minerology. Therefore, site-specific potash recommendation can attain through having the background knowledge about sorption capacities of particular soil. Different K pools in the soil can provide indication of K availability, and better availability of K for the plant is interpreted with quantification of these pools. The physicochemical characteristics of soils, and type and ratio of clay minerals plays a key role in K sorption and fixation on exchange sites, ultimately its availability to plants. Indeed, clay mineralogy and degree of soil weathering may have a substantial role in evaluating and optimizing K availability for plant. Furthermore, as a result of intensive agricultural practices along with introduction of high yielding fertilizer responsive cultivars without regarding significance of potash addition; our soils are prone to K deficiency. Therefore, for the better crop production on the sustainable basis, there is a need to optimize K fertilization in distinct agroecosystems of country.

In this book, we summarize the current state of knowledge of the consequences of K in the soil, plant, human and its relevance to arable farming in Pakistan. As far as possible, highlight the current status of potash in the context of agricultural sustainability, potential to promote balanced use of fertilizer, and strategies for more effective transfer of information at ground level to pursue sustainable productivity under local farming conditions. Furthermore, future research directions have been identified to promote research and development of potash fertilization in Pakistan. It can be beneficial to understand the overall biological hierarchy of K and possible potential of integrated potash management for higher yield and quality of agriculture production.

References

Akhtar ME, Saleem MT, Stauffer MD (2003) Potassium in Pakistan agriculture. Pakistan Agric Res Council, Islamabad 80

Amer MA, Attalah MZ, Ibrahim MF, El-Taweel F (2004) Effect of potassium fertilizer on mineral contents and some biological components of ten sugar beet varieties. J Agric Sci Mansoura Univ 29(6):2963–2978

Calabrese, A., Gibby, C., Meinke, B., Revilla, M.K.F. and Titchenal, A. (2019). Human nutrition. http://pressbooks.library.ryerson.ca/vitalsign/

Chen Z, Zhou M, Newman IA, Mendham NJ, Zhang G, Shabala S (2007) Potassium and sodium relations in salinised barley tissues as a basis of differential salt tolerance. Funct Plant Biol 34(2):150–162

Dobermann A, Witt C, Dawe D (2004) Increasing productivity of intensive rice systems through site-specific nutrient management. Science Publishers and International Rice Research Institute (IRRI), Enfield (NH) and Los Banos (Philippines)

Dunn D, Stevens G (2005) Rice potassium nutrition. Res Progress Better Crop 89(1):15–16

Etemadi AMA (2000) Effect of nitrogen and potassium fertilization on yield and quality of sugar beet in green house. Sugar beet. J Agric Res 78:759–767

Gul M, Wakeel A, Steffens D, Lindberg S (2019) Potassium-induced decrease in cytosolic Na^+ alleviates deleterious effects of salt stress on wheat (*Triticum aestivum* L.). Plant Biol 21(5):825–831

Holland JE, Cammarano D, Fitzgerald GJ, Perry EM, Poile G, Conyers MK (2019) Proximal fluorescence sensing of potassium responsive crops to develop improved predictions of biomass, yield and grain quality of wheat and barley. Precision Agric 20:379–397

Koné B, Amadji GL, Aliou S, Diatta S, Akakpo C (2011) Nutrient constraint and yield potential of rice on upland soil in the south of Dahomey gap of West Africa. Arch Agronomy Soil Sci 57(7):763–774

Marschner H (2012) Marschner's mineral nutrition of higher plants. Academic Press, London, UK

Mengel K, Kirkby EA (1982) Principles of plant nutrition. International Potash Institute, Bern, Switzerland

Mengel K, Kirkby EA (2001) Principles of plant nutrition. Kluwer Academic Publishers, Dordrecht, NL

Mubarak MU, Zahir M, Ahmad S, Wakeel A (2016) Sugar beet yield and industrial sugar contents improved by potassium fertilization under scarce and adequate moisture conditions. J Integrative Agric 15(11):2620–2626

Mussgnug F, Becker M, Son TT, Buresh RJ, Vlek PLG (2006) Yield gaps and nutrient balances in intensive, rice-based cropping systems on degraded soils in the red River Delta of Vietnam. Field Crops Res 98(2–3):127–140

Ouda MMS (2002) Effect of nitrogen and Sulphur fertilizers levels on sugar beet in newly cultivated sandy soil. Zagazig J Agric Res 29:33–50

Rehm G, Schmitt M (2002) Potassium for crop production. http://www.extension.umn.edu/agriculture/nutrient management/potassium/potassium-for-crop-production/

Römheld V, Kirkby EA (2010) Research on potassium in agriculture: needs and prospects. Plant Soil 335:155–180

Shahzad AN, Rizwan M, Asghar MG, Qureshi MK, Bukhari SAH, Kiran A, Wakeel A (2019) Early maturing Bt cotton requires more potassium fertilizer under water deficiency to augment seed-cotton yield but not lint quality. Sci Rep 9(1):1–10

Wakeel A (2013) Potassium–sodium interactions in soil and plant under saline-sodic conditions. J Plant Nutr Soil Sci 176(3):344–354

Wakeel A, Hassan A, Aziz T, Iqbal M (2002) Effect of different potassium levels and soil texture on growth and nutrient uptake of maize. Pakistan J Agric Scis 39:99–103

Wang Y, Wu WH (2017) Regulation of potassium transport and signaling in plants. Curr Opin Plant Biol 39:123–128

Wang M, Zheng Q, Shen Q, Guo S (2013) The critical role of potassium in plant stress response. Int J Mol Sci 14:7370–7390

Weaver CM (2013) Potassium and health. Adv Nutr 4(3):368S–377S. https://doi.org/10.3945/an.112.003533

2 Potassium Dynamics in Soils

Potassium (K) is a macronutrient required by plants in larger amounts such as nitrogen, however, different from all macronutrients because of its specific retention in the soils. It is a structural part of most of the soil minerals and its availability to plants is not directly proportional to the total amount of K in the soils. Potassium availability can be understood only by understanding the K content in soil, forms of K, soil mineralogy, and weathering status of the soils.

2.1 Potassium Content in Soil

Currently, the world's most important potash deposits are located in (1) the Elk Point, Maritimes, Paradox, and Salado Basins in North America; (2) the Zechstein, Pripyat, Solikamsk, and Pricaspian Basins in Europe; (3) the Danakil and Lower Congo Basins in Africa; (4) the Dead Sea, Khorat, Sakon Nakon, Qaidam, and Central Asia Basins in Asia; and (5) the Amazonas, Sergipe, and Neuquén Basins in South America (Fig. 2.1). Major potash producer countries are Canada, followed by Russian Federation, Belarus, Germany, Israel, Jordan, China, USA, Brazil, Chile, Spain, and UK (Jasinski 2011). In cultivated areas, K concentration may vary from 0.01% to 4%, with an average calculated value of 1.36% (Jakovljević et al. 2003). In mineral soils, the average concentration of K is ~1.4%, with a broad range of 0.01% to 3.7%. The total soil K contents frequently exceed 20,000 mg kg^{-1} in most soils (Rehm and Schmitt 2002). In rocks and several soils, K is a major constituent of the primary rock-forming minerals, for instance, alkali feldspar group, namely microcline and orthoclase. Mica group, specifically biotite and muscovite and secondary minerals such as illite are also a rich source of K (Jakovljević et al. 2003).

The concentration of soil available K varies with the degree of weathering and the soil parent material which has direct impact on its supply (Table 2.1). In soils, K exists in four major forms, namely structural K, exchangeable (readily available), non-exchangeable or fixed (slowly available), and readily available solution K

A. Wakeel, M. Ishfaq, *Potash Use and Dynamics in Agriculture*,
https://doi.org/10.1007/978-981-16-6883-8_2

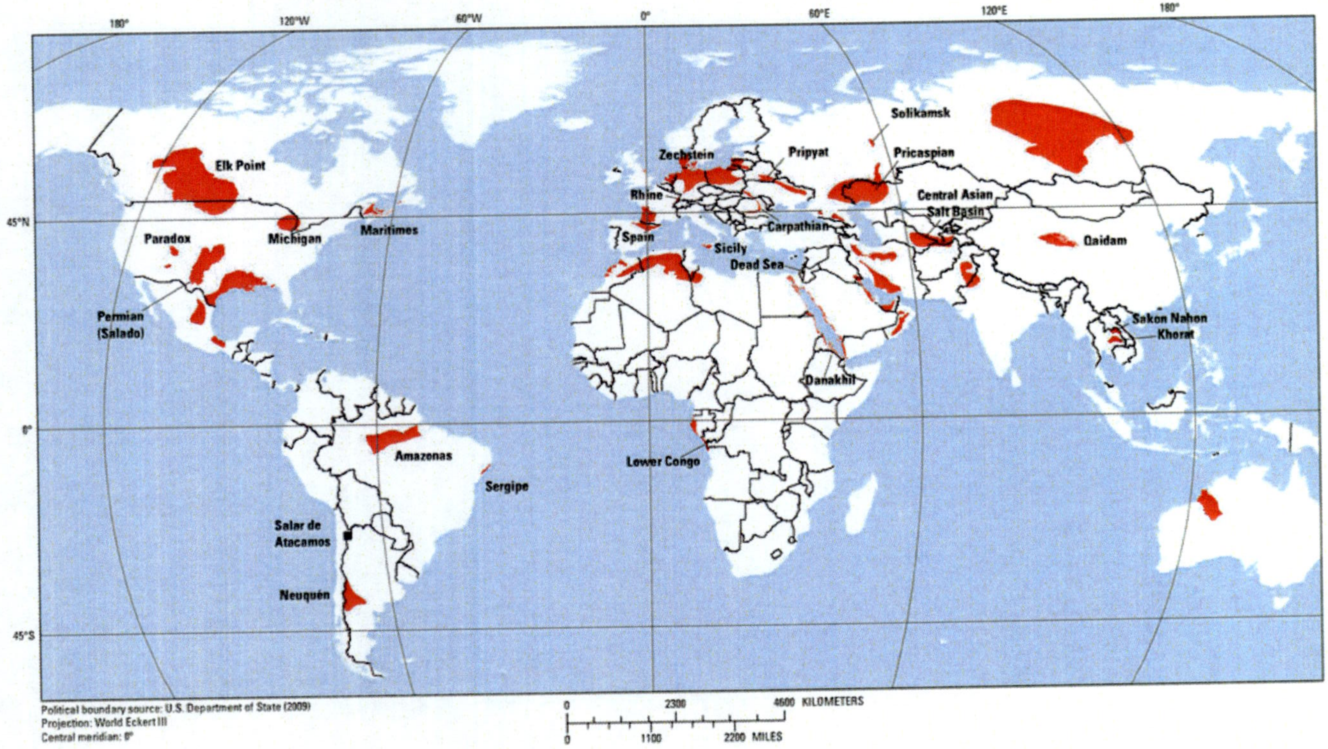

Fig. 2.1 World map of significant potash-bearing marine evaporite basins. Basins are in red, and the major potash-producing salar in Chile is depicted by a black square (adapted from Cocker et al. 2016). For more detail, interactive geographic information system-based maps and data are available at: https://pubs.usgs.gov/sir/2010/5090/s/pdf/sir2010-5090-S.pdf

Table 2.1 Total potassium concentration in some soil orders

Soil order	K concentration (mg g^{-1} soil)
Entisols	26.3 ± 0.6
Spodosols	24.4 ± 0.5
Alfisol	11.7 ± 0.6
Mollisol	11.7 ± 0.6

Source: Helmke PA (2000). In Sumner ME (ed), Handbook of soil science. CRC Press, London, p B3–B24

(Rehm and Schmitt 2002). It has been revealed that more than 90% of soil K is present in the form of primary minerals, while the amount of plant available K in the soil solution is only about 0.2%, and it is the principle source of K for plant uptake (Fig. 2.3). The fixed form of K is called as structural, mineral, native, matrix or inert K. In many soils, the total K or the bulk of it is structural K is not directly available to plants. It has been found that nearby 90–98% K in the soil is in the unavailable form depending on soil type. This K is released through the weathering process, nevertheless this process is so slow that it cannot meet crop requirements. However, fixed K contents influence the concentration of soluble K in soil solution (Jakovljević et al. 2003).

The second type of K present in the soil is the exchangeable K; which is present at the exchange sites of various soil colloid complexes. This K is held by different bond strengths at the non-specific adsorption sites. These sites may include the planer and edge positions of soil clay minerals or the negative charge sites which are developed by carboxylic and phenolic groups of humus colloids. The amount of K held at exchangeable sites of soil clay minerals depends upon the kinetic and thermodynamic factors (Parfitt 1992), for instance, the K concentration at exchangeable site in relation to other exchangeable cation, and the nature of the soil surface. Approximately 90% of available K is present in exchangeable K, whereas the proportion of exchangeable to total K is <2%. It is stated that exchangeable K has a strong link with CEC, pH, calcium carbonate, and organic carbon (Baruah et al. 1991). For example, aridisols and inceptisols have comparatively higher K rich mica (Srinivasarao et al. 2007), and K present in illite have less mobility as these minerals have narrow interlayer space which are K selective and ultimately decrease its desorption (Sparks and Huang 1985; Sparks 1987). Furthermore, sub-surface soils have less exchangeable K than surface soils and higher concentration of exchangeable K in the soil represent a good indication of ability of soil to supply K. Some exchangeable K may fix between layers of clay minerals and may be converted into slowly available K (Rehm and Schmitt 2002).

The third form of K is the non-exchangeable, fixed or slowly available, which is trapped between the clay mineral layers, and plants cannot use this K fraction during a single season. Nevertheless, this K can act as a reservoir for exchangeable K that is readily available (Rehm and Schmitt 2002). This form of K is held between the adjacent tetrahedral layers of di- and tri-octahedral vermiculites and micas or between the chloritized vermiculite (Sparks and Huang 1985; Sparks 1987). The fixation of K takes place due to the increase in the number of binding forces between

K and clay surfaces than the hydration forces which are present between the individual ions of K (Sparks 1987), making the release of K easy due to physical trapping. The conversion of non-exchangeable K into the exchangeable K occurs when the solution K and exchangeable K level are decreased due to K removal by crops, losses through leaching or through enhanced activates of soil microbes (Sparks 2000).

The non-exchangeable K is moderately to sparingly available to plants. The higher concentration of non-exchangeable K is noticed in sub-surface soils than surface soil. Furthermore, the soils enriched with illit clay minerals have high non-exchangeable K in the sub-surface because of the presence of clay and silt which have ability to fix K (Sharma et al. 2009). This K fixation depends on the amount of K released from clay minerals, particle size, and their distribution/quantity and type of clay minerals, because these properties have strong correlation with clay and silt proportion of soil. The quantity of K fixed in soil and type of clay minerals rich in K can influence the release of K to plants (Rao et al. 2001).

The fourth form of K is in the solution form which is readily available to plant. Readily available K may also be present at the exchangeable site of clay minerals which has been discussed earlier. The quantity of K in soil solution depends upon absorption through the plants and also involves the soil minerals from which K is released. When the plants dissolve K present in soil solution, the concentration of K in soil solution drops, which is balanced after the release of K from exchangeable sites of clay minerals. This form of K mostly faces a leaching problem in soil. In the soil solution, the amount of K is not constant and ranges from 3 to 30 mg kg^{-1} for normal soils (Haby et al. 1990). The solution K level is mediated by kinetic reactions, soil moisture contents, and the equilibrium that occur between the forms of soil K, the concentrations of divalent cations in solution and on the exchanger phase (Sparks 2000). The amount of soil solution K is also affected by the buffering capacity of soil clay minerals. For example, well-buffered vermiculates and micas can hold the solution K level very firmly, while that of buffered allophonic clay mineral do not maintain the soil solution K level (Parfitt 1992). Additionally, slowly available K, exchangeable K, and water-soluble K are interconvertible (Fig. 2.3), and this conversion is mediated by various factors, for instance, K supply, soil moisture, soil temperature, and the removal of K via plant roots.

The removal of K by the plant is facilitated by K concentration in soil solution, K availability index, organic matter, clay minerology, cation exchange capacity, calcium carbonate, and other properties. The knowledge regarding the variation in K adsorption, differences among various soil types and the equilibration between K intensity and the phases of the quantity is valuable for precise and site-specific potash recommendation. Furthermore, the K contents in the soil may increase and decrease depending on environmental and soil conditions. The indigenous soil K contents are depleting with time owing to intensive cultivation of crops with less potash fertilization and other crop management practices, particularly the removal of crop residues from the field.

2.2 Soil Mineralogy and Potassium Dynamics

Soil clay mineralogy has great consequence in soil nutrient supplying capacity and overall nutritional status of a particular soil. It is exposed that the mineralogical composition of the soil may largely impact the K dynamics within the soil (Surapaneni et al. 2002), and this relationship between the K forms and clay mineralogy can be used to predict soil K fertility, and ultimately the potash fertilizer recommendation. Clay mineralogy in the soil influences the fixed K which is not readily available fraction of K in the soil. For instance, when the soil becomes dry, montmorillonite clays fix K because K is trapped in clay mineral layers. When the soil becomes wet, this trapped K release. In contrast, illite dominant soils, the K is fixed between the layers of clay minerals when the soil is dry, but releases only small amount of K when soil becomes wet. Furthermore, clay minerals such as smectite and kaolinite easily release all of their adsorbed K than do illite and vermiculite (Mengel and Uhlenbecker 1993). Therefore, the age of soil developed from these clay minerals determines the extent of weathering are responsible to regulate the soil K dynamics (Fig. 2.2).

The major share of K exists in micas and feldspars clay minerals which contain 6–9% and 3.5–12% of the total soil K and such 2:1 type of clay mineral revealed as major sources of non-exchangeable K (Moritsuka et al. 2004). However, the plants are unable to use this crystalline-insoluble K present in these minerals. Weathering of these rocks result in the release of K; some of this K moves towards the readily available pool and some towards slowly available pool. Thus, this K fixation in illite clays without release causes serious problems for management of potash fertilizer for crop production (Rehm and Schmitt 2002). The other clay minerals which bear K are primary aluminosilicates such as micas-muscovite, biotite and secondary

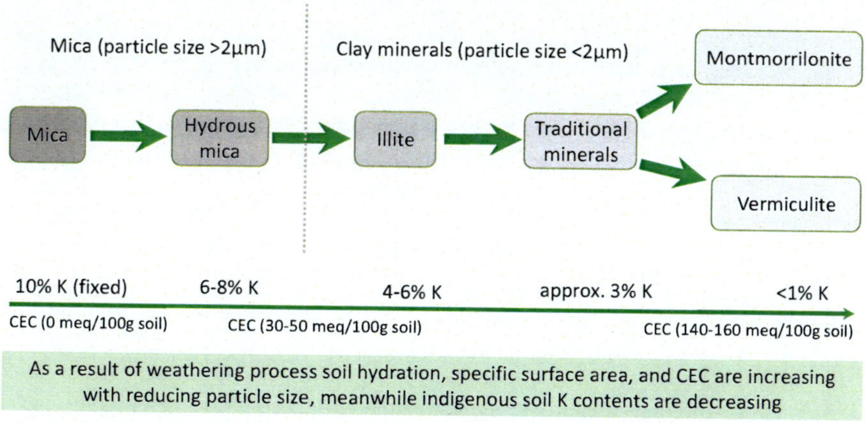

Fig. 2.2 Schematic diagram of the soil weathering process in relation to potash availability. With the passage of weathering process; soil surface area expands having smaller particle size, whereas indigenous K content reduces

Table 2.2 Representative K concentrations (mg kg^{-1} soil) in soil fractions related to dominating clay types

K fraction	Kaolinitic ($n = 26$)	2:1 type clay ($n = 23$)	Mixture ($n = 53$)
Exchangeable	45	183	224
Water-soluble	2	4	5
Total	3340	15,780	8920

Source: From Brady NC, Weil RR (1999) The nature and properties of soils. 12th edition. Prentice-Hall, Englewood Cliffs, NJ

aluminosilicates, namely phyllosilicates. The formation of these phyllosilicates occurs through the weathering of primary minerals or from the secondary silicates after adequate rainfall (Essington 2004). The K availability to plants is very easy from phyllosilicates and its associated clay minerals. A majority of clay particles in soil have particle size of <0.002 mm, and are made from the secondary clay minerals.

The exchangeable K may bind on non-specific adsorption sites of clay minerals, at the carboxylic and phenolic groups of soil and also on soil organic matter. The absorption of the K^+ takes place non-specifically in hydrated form. These non-specific adsorption sites are present on the edge "e" and planar "p" position of clay minerals, and aluminum hydroxides and amorphous iron (Sparks 2003). The lattice or structural K is covalently bonded with the crystalline structure of K enriched primary silicate minerals such as feldspars, namely microcline, orthoclase showing framework structure and micas, namely muscovite biotite showing layered structure. The micas consist of ~8% of soil K. These types of clay minerals are present in the coarser fractions of soils (e.g. sand and silt) and the contained K in these minerals is release through the long-term weathering process. However, this K is not practically available to plants. The typical process of weathering proceeds in following order: (1) muscovite, (2) hydro muscovite, (3) illite, (4) mixed clay minerals, e.g. kaolinite-illit or illit-montmorillonite (Fig. 2.2). In this weathering process, the K ion is substituted by hydroksyanion H_3O^+, and thus the next product contains less K (Pracz 2003).

Although K is found in large quantities in world soil, the immediate plant available K is non-sufficient to meet the requirement of a specific crop. Beside this, the added K fertilizer may adsorb on soil colloids, may fix in clay minerals, or can leach to the underground water, especially in sandy soils. It has been found that ~57% of the applied K can be adsorbed by soil colloid depending on the type and amount of soil clay minerals (Pal et al. 1999), for example, illite and vermiculite have higher K adsorption specificity. In contrast, some studies have quantified that the rate of K adsorption on illite and vermiculite clay minerals is slower than that recorded on kaolinite and montmorillonite. The particle size and chemical composition of the clay minerals affects considerably the release of K from clay minerals (Huang 2005) (Table 2.2). Furthermore, the fixation process of K is relatively fast, whereas the release of fixed K is very slow due to the strong binding force between K and clay minerals (Öborn et al. 2005). Whether a soil fixes or releases K highly depends on the K concentration in the soil solution (Schneider et al. 2013).

2.3 Nature of Minerals in Pakistani Soils

The soils of Pakistan are composed of the various types of clay minerals such as muscovite, biotite, mica, smectite and vermiculite, feldspar, quarts, chlorite, montmorillonite, kaolinite, and illite. Among these clay minerals, both mica and illite are dominent in the soils of Pakistan, and it determines the chemistry of soil solution, and thus the fertilizer behavior (Fig. 2.3). The silt and sand fractions of these soils consist of primary clay minerals such as mica, feldspar, and quarts; mica and feldspar present in large fractions and are ores of K. The clay fractions consist of secondary clay minerals, namely vermiculite, chlorite, montmorillonite, kaolinite, illite, montmorillonite, and kaolinite being most abundant. Although the soils of Pakistan also have chlorite and kaolinite clay minerals which contain K, but their significance as potash nutrition is low due to their low nutrient retention ability. Indeed, the poor drainage and semi-arid environmental conditions may cause the further weathering of these both clay minerals to saponitic smectites.

The alluvial soils of Pakistan have both clay minerals, namely mica and illite; crucial source of K nutrient. Mica is most dominate K bearing clay mineral in Pakistan which is formed by slow weathering of Himalayan rocks. The K contents of mica in clay, silt, and sand are 97%, 93%, and 70%, respectively (Ali et al. 2006). During weathering, mica releases the K ions into soil solution. If the process of weathering proceeds enough, mica is converted into expandable illite, also known as hydrated mica, after the entry of water in interlayer. The hydrous mica in Pakistani soils is of two types namely biotite mica and muscovite. Among both these types, the biotite mica is more important form fertility point of view as it releases the K, 100 times more than muscovite in soil solution (Ali et al. 2006).

Like other clay minerals, smectite and vermiculite are found in very low quantity in most of the soils in Pakistan, but these clay minerals play a crucial role in the fixation and release of K fertilizers due to their shrinking and swelling properties and higher nutrient exchange capacity. The formation of vermiculite clay minerals takes place after intensive weathering of illite, accomplished with the release of K. In soils, where the vermiculite is a dominant clay mineral, the phenomenon of fixation may occur due to high nutrient exchange capacity and high adsorption selectivity of vermiculites for ammonium and K than other nutrients, which may reduce the fertilizer use efficiency. However, the vermiculite clay minerals are rarely developed in the Pakistani soil except in Murree, Wazirabad, Gujranwala, and Satgarh soil series (Ali et al. 2006).

In Pakistani soils, the smectite clay mineral can be divided into beidellite and montmorillonite. Under wet conditions, both of these clay minerals expand, while both shrink on drying. Regarding the K fixation, the behavior of beidellite clay mineral is similar to vermiculite, while the K fixation capacity of montmorillonite is 50 times less than beidellite. Thus, in Pakistani soils, the occurrence of beidellite and

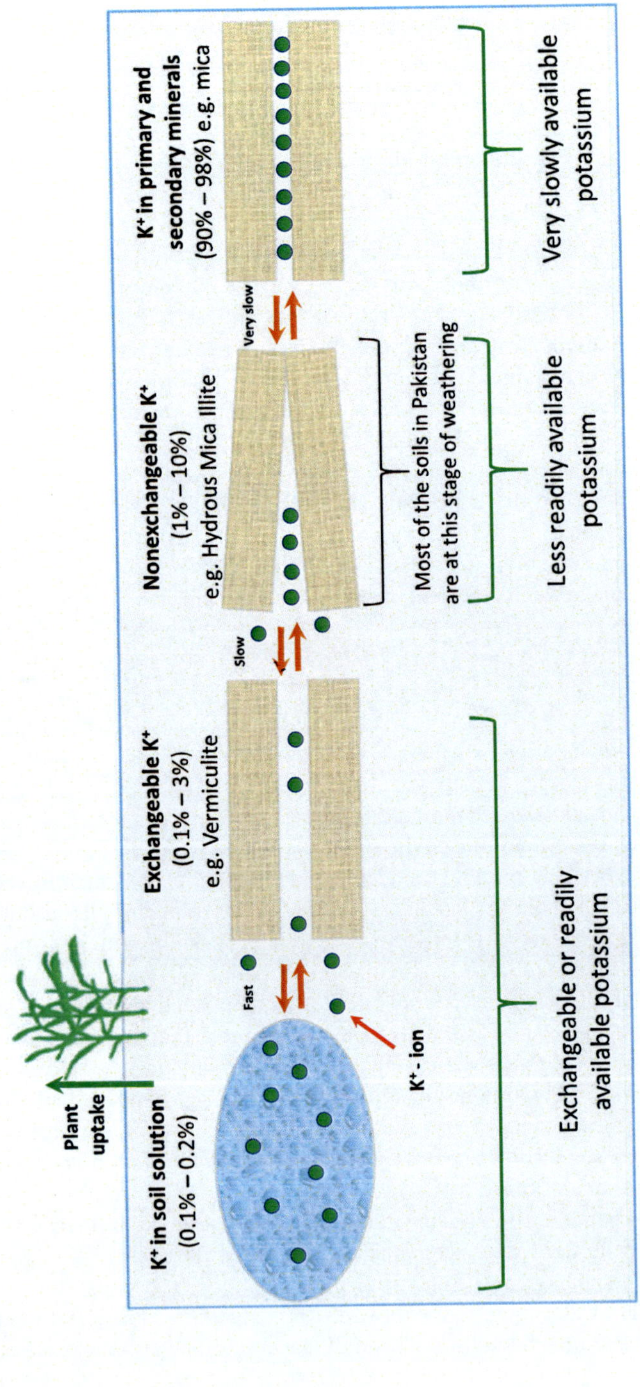

Fig. 2.3 The major fraction of potassium in soils is in the fixed form which is not readily available for plants. Most dominant clay minerals in agricultural soils of Pakistan are illite/hydrous mica; however, vermiculite has also been reported in small areas

vermiculite is highly related to K deficiency, while the occurrence of illite and montmorillonite to optimum K supply status. A soil can be regarded as best which contain montmorillonite and both illite/mica, because the K released from both these clay minerals can be held as an exchangeable nutrient by montmorillonite clay mineral with less fixation but more easy availability to crop roots (Ali et al. 2006).

2.4 Plant Available Potassium

The amount of K present in the soil is evaluated by a number of soil factors such as its uptake by plant, degree of weathering such as the amount and type of clay, the addition of fertilizer, leaching losses, and the fixation process/K release that occur in the soil. As mentioned earlier, the plants are unable to take up most of the soil indigenous K because it is present in structural or fixed form. Plants can take up only those fractions of the K which are present in the dissolved form (<100 mg kg^{-1}) or exchangeable form (Rehm and Schmitt 2002). Furthermore, the supply of K to plant is directly controlled by the K concentration in soil solution. Additionally, some soils with low plant available K can maintain the high levels of solution K which fulfill the crop requirements and hence the outer K input did not impact crop growth as the needs are fulfilled by K already in soil.

In Pakistan, K content in soils has traditionally been considered adequate for normal plant growth as they are developed from mica type of clay minerals; but in most soils, most of this K is not necessarily available to plants to complete their life cycle. In such soils, K is bound to the clay mineral which do not release the K at the rates which are required for optimum crop harvest. In an early study, it was reported that most of soils in Pakistan have less than 150 mg kg^{-1} of exchangeable K (Bajwa and Rehman 1996), and it is considered deficient to fulfill the requirements of most of the crops. Furthermore, the extensive cropping with high yielding cultivars accomplished with no potash fertilizer in Pakistani soils have seriously dropped soil indigenous K level. More than 30% of soils in Pakistan require potash fertilizer because these soils are K deficient (Akhtar and Dixon 2009). However, the application rate is also very low and according to the most recent report only 16% farmers in cotton growing areas of Pakistan use potash containing fertilizers while 44% soils are revealed K deficient (Wakeel and Ishfaq 2016).

In intention to assess the changing pattern of soil K levels in Pakistan, data from soil samples analyzed for advisory purpose to farmers from two particular years, namely 1990 and 2005 were compared (Table 2.3). The Punjab province of Pakistan was categorized into four agroecological zones, namely Rainfed Zone comprises of northern parts of the province; Rice Zone consists of irrigated north-eastern parts of the province; Central Zone comprises of central irrigated districts of the province and cotton Zone falls in southern parts of the province. It was observed that within 15 years; soil available K declined 7–49% in different agroecological zones of Punjab province. Such dramatic depletion of K is alarming, and soils are prone to additional K deficiency if potash fertilizers will not be oriented in farmer input practices.

Table 2.3 Soil potassium level (mg kg^{-1}) in various crop zones of Punjab–Pakistan by 1990 to 2005 (analyzed by Soil Fertility Research Institute, Punjab)

	1990		2005		
Zone	No. of samples	Mean value	No. of samples	Mean value	Percent decrease
Rainfed	1500	114	1550	106	7
Rice	1400	176	1645	132	33
Central	2670	295	3200	198	49
Cotton	1658	210	2280	165	27

Source: Ahmad and Mian (2009)

References

Ahmad Z, Mian SM (2009) Increasing crop productivity through judicious use of potash in Pakistan. Proceedings IPI-OUAT-IPNI International Symposium

Akhtar MS, Dixon JB (2009) Mineralogical characteristics and potassium quantity/intensity relation in three Indus River Basin soils. Asian J Chem 21(5):3427

Ali, K, Khan, IA, Azeem, M, Ahmed, Z, Abbas, S (2006) Clay minerals and fertilizer management. http://www.dawn.com/news/189092/clay-minerals-fertilizer management

Bajwa MI, Rehman F (1996) Soil and fertilizer potassium. Soil science. National Book Foundation, Islamabad, p 319

Baruah HC, Bora DK, Baruah TC, Nath AK (1991) Fixation of potassium in three major soil orders of Assam. J Pot Res 7(3):170–175

Brady NC, Weil RR (1999) The nature and properties of soils, 9th edn. Macmillan Publishing Company New York

Cocker, M.D., Orris, G.J. and Wynn, J. 2016. U.S. Geological Survey assessment of global potash production and resources—A significant advancement for global development and a sustainable future. The Geological Society of America Special Paper 520.

Essington ME (2004) Soil and water chemistry, an integrated approach. RC Press, London

Haby VA, Russelle MP, Skogley EO (1990) Testing soils for potassium, calcium, and magnesium. Soil Testing Plant Anal 3:181–227

Helmke PA (2000). In: Sumner ME (eds) Handbook of Soil Science, CRC Press

Huang PM (2005) Chemistry of potassium in soils. In: Tabatabai MA, Sparks DL (eds) Chemical processes in soils, vol 8. Soil Science Society of America, pp 227–292

Jakovljević MD, Kostić NM, Antić-Mladenović SB (2003) The availability of base elements (Ca, mg, Na, K) in some important soil types in Serbia. Zb Matice srp prir nauke 104:11–21

Jasinski, S.M. 2011. Potash: Mineral Commodity Summaries 2010: U.S. Geological Survey, pp. 122–123. http://minerals.usgs.gov/minerals/pubs/commodity/potash/mcs-2011-potas.pdf.

Mengel K, Uhlenbecker K (1993) Determination of available interlayer potassium and its uptake by ryegrass. Soil Sci Soc Am J 57(3):761–766

Moritsuka N, Yanai J, Kosaki T (2004) Possible processes releasing nonexchangeable potassium from the rhizosphere of maize. Plant Soil 258(1):261–268

Öborn I, Andrist-Rangel Y, Askekaard M, Grant CA, Watson CA, Edwards AC (2005) Critical aspects of potassium management in agricultural systems. Soil Use Manag 21:102–112

Pal Y, Gilkes RJ, Wong MTF (1999) The forms of potassium and potassium adsorption in some virgin soils from South-Western Australia. Soil Res 37(4):695–710

Parfitt RL (1992) Potassium-calcium exchange in some New Zealand soils. Soil Res 30(2):145–158

Pracz J (2003) Podstawy mineralogii. SGGW, Warszawa

Rao CS, Rupa TR, Rao AS, Bansal SK (2001) Subsoil potassium availability in twenty-two benchmark soil series of India. Commun Soil Sci Plant Anal 32(5-6):863–876

Rehm, G. and Schmitt, M. 2002. Potassium for crop production. http://www.extension.umn.edu/agriculture/nutrient-management/potassium/potassium-for-crop-production/

Schneider A, Tesileanu R, Charles R, Sinaj S (2013) Kinetics of soil potassium sorptiondesorption and fixation. Commun Soil Sci Plant Anal 44:837–849

Sharma A, Jalali VK, Arya VM, Pradeep R (2009) Distribution of various forms of potassium in soils representing intermediate zone of Jammu region. J Indian Soc Soil Sci 57(2):205–207

Sparks DL (1987) Potassium dynamics in soils. Adv Soil Sci 6:1–63

Sparks DL, Huang PM (1985) Physical chemistry of soil potassium. In: Munson RD (ed) Potassium in agriculture. Soil Science Society of America, Madison, pp 201–276

Sparks DL (2000) Bioavailability of soil potassium. In: Sumner ME (ed) Handbook of soil science. CRC Press, Boca Raton, pp 38–52

Sparks DL (2003) Environmental soil chemistry. Elsevier

Srinivasarao C, Vittal KPR, Tiwari KN, Gajbhiye PN, Kundu S (2007) Categorisation of soils based on potassium reserves and production systems: implications in K management. Soil Res 45(6):438–447

Surapaneni A, Palmer AS, Tillman RW, Kirkman JH, Gregg PEH (2002) The mineralogy and potassium supplying power of some loessial and related soils of New Zealand. Geoderma 110 (3-4):191–204

Wakeel A, Ishfaq M (2016) Promoting precise and balanced use of fertilizers in Pakistan at farm-gate level. E-ifc 47:30–35

Potassium in Plants

3

3.1 History as an Essential Element

Potassium was discovered in 1807 by Sir Humphry Davy in England. The letter K, which is used to symbolize potassium, has been derived from the Latin word "*Kalium*." Prior to the industrialization, soap was manufactured by burning wood and other organic matter in the pot. The ashes obtained after burning of organic matter and wood were washed or purified, allowing the water to evaporate, in this fashion, leaving residues of K salts. At that time people called these collected residues as "*pot ashes or potash*" and they were further boiled with animal fat to make soap. Afterward, a botanist named Samuel William Jackson, analyzed the ashes in 1868, after burning of plants. Jackson distinguished K and numerous other minerals in these ashes, and his findings lead toward the use of K fertilizer to boost crop growth and yield. The first patent about an improved potash production method was issued by the government of the USA.

The term "*essential mineral element or nutrient*" was at first suggested by Arnon and Stout in 1939. They proposed that three criteria for an element must be met to be considered essential for plant, and K fulfills all the below mentioned requirements.

1. A given plant must be unable to complete its life cycle in the absence of the element.
2. The function of the element must not be replaceable by another element.
3. The element must be directly involved in plant metabolism or a structural component.

Globally, China is reported as a leading consumer of potash, followed by USA and India while Canada is the largest producer and exporter of potash fertilizer with annual production of about 8 million tons followed by Belarus, Russia, and Germany. However, the prices of potash fertilizers have been increased since last few years, which is forcing the mining industry to found alternate sources of raw

A. Wakeel, M. Ishfaq, *Potash Use and Dynamics in Agriculture*,
https://doi.org/10.1007/978-981-16-6883-8_3

material for potash production. Although K is the seventh most abundant element found in the Earth's crust, nonetheless, as mentioned earlier in the soil only 1–2% is present in plant available form, and the rest is fixed within minerals. The majority of K is mined (some time as deep as 3000 feet) from the underground deposits same like the coal mining. The solution mining is also practiced in order to dissolve the ore. This solution containing the dissolved ore can evaporate, thus leaving the K salts behind. Few fractions of K salt are also obtained by evaporation of water from the natural salt lakes/sea around the globe, for instance, the Dead Sea in Jorden and Israel and Great Salt Lake in Utah.

Worldwide, potash is marketed in both liquid and granular form or is generally mixed. On fertilizer labels, water-soluble K is symbolized as K_2O which does not exist in soil, plant, or fertilizer. Plant uptake it in the ionic form (K^+) after dissolution in water. About 95% of the K fertilizer is marketed as potassium chloride/muriate of potash (MOP). However, potassium nitrate and potassium sulfate are the other sources of potash commercially available which are preferable for salt sensitive crops, nonetheless they are expensive (see Chap. 8).

3.2 Functions in Plants

Potassium is a key element in plant nutrition and physiology. It activates over 60 enzymes, promotes photosynthesis activity, regulates stomatal opening, promotes the translocation of photosynthates from the source to sink organs, and detoxify reactive oxygen species (ROS). It is termed as *"quality element"* as its adequate supply optimizes yield as well as quality traits of agricultural commodities (see Chap. 6 for detail). In addition, it protects the plant from the adverse effects of both biotic and abiotic stresses which is comprehensively highlighted in Chap. 8, and improves the utilization of nitrogen, consequently, reduces environmental footprints (see chap. 11).

3.2.1 Enzyme Activation

In plants, enzymes act as catalysts to proceed various chemical processes. Interestingly, K is the activator of enzymes participating in energy metabolism, protein synthesis, and solute transport. Indeed, K is bound electrostatically at the surfaces of enzymes and thus alters the conformation of the molecules of enzymes, thus subjecting the suitable active sites for the chemical reaction (Fig. 3.1). Thus, adequate K supply is crucial for an efficient operation of various plant body processes (Taiz and Zeiger 2010). The concentration of K which is existing in plant cells principally regulates the rate of any chemical reaction and the number of activated enzymes. Consequently, that rate at which K enters the cell principally determined the rate of certain chemical reaction (Van Brunt and Sultenfuss 1998).

The deficiency of K in plants also trigger the synthesis of protein by affecting the uptake of nitrate (see Chap. 11 for detail). The nitrate, amides, and amino acids (well

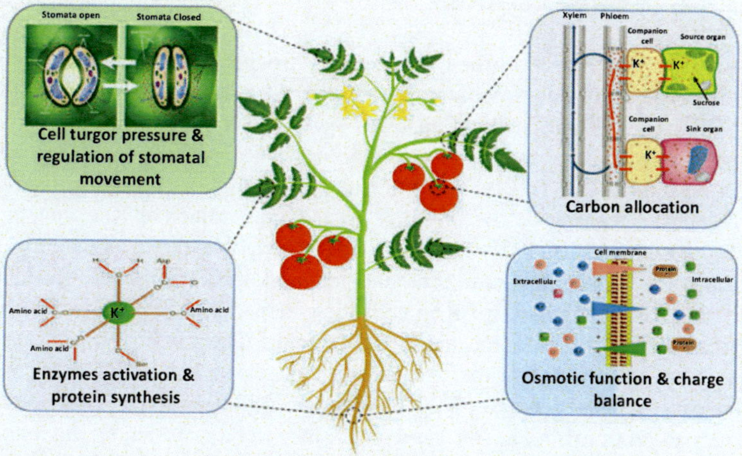

Fig. 3.1 Functions of K in numerous physiological and biochemical processes including enzymes activation, charge balance, carbon allocation from source to sink organs by governing sucrose phloem loading as well as osmotic function which promote to maintain cell turgor pressure and stomatal movement

know protein precursors) can accumulate in plant when the supply of K is inadequate. Plants possibly sense the available level of NO_3^- in soils and adjust accordingly to uptake and transport from root-to-shoot to balance the distribution of these ions between plant organs (Raddatz et al. 2020). The translation and transcription (vital steps of protein synthesis) are negatively impacted without an adequate supply of K in crop plants. In a study on sweet sorghum, the combined application of N and K enhanced growth attributes and juice/extract (Almodares et al. 2008) which can be associated with their possible role in protein synthesis.

The involvement of K in binding the ribosomes and tRNA has been revealed in many research studies; however, to date, it is not well characterized which specific enzyme of ribosomal site or translation activate with K availability (Mengel 2007). Potassium is also believed to trigger the biosynthesis of a co-enzyme (NADPH) during the photosynthesis activity. Much earlier, Hawker et al. (1974) pointed K as an important element for the activation of enzymes which involve in the synthesis of starch. With an increment of K, the starch level improves, meanwhile K deficiency causes the accumulation of soluble carbohydrates. Indeed, the increase in N compounds, the accumulation of soluble carbohydrates, and the reduction in the synthesis of starch in K deficient plants are due to reduced activity of various enzymes, for example, phosphofructokinase and pyruvate kinase, as these enzymes demand K for their adequate functionality. In addition, K is positioned as an octahedron with oxygen atoms at six corners in dialkyl-glycine decarboxylase enzyme. These atoms of oxygen possess one water molecule, three amino acyl molecules, and O of the hydroxyl groups of aspartate and serine (Miller 1993).

3.2.2 Charge Balance

For the transfer of ions, the buildup of electrochemical gradient is essential between the cytosol and outer medium. Marschner (2012) acknowledged that K, apart from numerous enzymes activation, also performs a fundamental role in the charge balancing and the transport of metabolites within plant body. The cell membranes of the plants are relatively permeable to K, thus highlighting its role in charge balancing owning to the presence of different selective K channels (Yamaguchi and Blumwald 2005). Furthermore, when the electrochemical potential is reduced in the cytosol as compared to the outer medium, the K transport towards cells ceases. Thus, with the transfer of K to cell, the electrochemical gradient develops and as a final it accomplishes the equilibrium point. The continuous release of H^+ ions towards apoplast from the cytosol through plasmalemma H^+ activity maintain the cytosol negative charge, thus keeping the negative potential of membranes from 120 to 200 mV. The movement of the water, nutrients, phytohormones, and metabolites via the xylem and phloem (Chrispeels et al. 1999), and organic acids and amino-N compounds through phloem (Pettigrew 2008; Wang et al. 2013; Hu et al. 2019) are directly or indirectly regulated by adequate potash supply. Mengel (2007) reported K concentration of 60–100 mM in the phloem because for phloem loading and transport, the K ions are very vital. For example, the glutamine and sucrose are apparently transport into the sieve cells by co-transport at elevated apoplastic pH. Thus, at unsatisfactory H^+ supply, the K^+ can facilitate the transloca-tion of crucial plant metabolites essential for growth and development by substituting H ions. Reduced K^+ supply lowers the translocation processes within plant.

Potassium also supports to regulate the pH around 7 and 7.5 (favorable for optimum reactions of many enzymes) by neutralizing organic anions and several additional metabolites within plant. For instance, RuBisCO (ribulose-1,5-bisphosphate carboxylase/oxygenase) is a crucial enzyme that catalyzes the assimi-lation of carbon dioxide (CO_2) during the process of photosynthesis. However, this enzyme is activated at high pH (8.3–8.6), accomplished with proper availability of magnesium (Mg). In this context, K supply play important role to maintain optimum pH gradient for enhanced activity of RuBisCO due to better membrane permeability (Schubert 2006).

3.2.3 Osmotic Functions

The osmotic functions are accomplished satisfactorily with adequate K^+ supply due to its easy transport through the biological membranes. The water uptake through root cells becomes easy when K^+ decrease the soil water potential. The transmission of K^+ from the xylem parenchyma cells into the xylem vessels lowers the water potential thus attracting more water through developing root pressure. The accumu-lation of the K^+ within the plant vacuoles not only act as a K^+ storage house but also as an essential osmoticum. Marschner (2012) reported that K^+ has indispensable function to adjust osmotic potential of vacuole, over and above, under water stress conditions. For the maintenance of tissue turgor, it is essential that the turgor of the vacuole must be controlled due to large size of vacuole.

Generally, the osmotic function within the plants is non-specific due to involvement of various organic and inorganic osmotica. However, the K^+ is considered as one of the predominant osmoticum which is transported to the fast-growing tissues at sufficient concentration (Mengel 2007), and thus play a vital part to maintain the turgor within the phloem and motor/guard cells. Similarly, the opening and closing of stomata in plants is generally regulated by adequate K^+ supply (Oosterhuis et al. 2014; Yang et al. 2016). The movement of K^+ towards guard cells causes K^+ accumulation within cells thus causing their swelling. These swelling forces the pores to open thus allowing the exchange of oxygen and CO_2 freely from the atmosphere (Fig. 3.1). Furthermore, under water deficit conditions, the guard cells release K^+, and the pores are firmly close to lessen the water loss to diminish drought stress effect on plants (Thomas and Thomas 2009).

3.3 Critical Concentrations of Potassium in Plants

Potassium concentrations in field crops vary widely depending on the site, year, fertilizer input, and plant species. In various plant species, its concentrations have been detected in the range of 0.4–4.3% (Askegaard et al. 2004). Furthermore, the most part of subcellular K^+ concentration is not well characterized with the exception of the cytosol and the vacuole. Plants maintain K concentration in the cytoplasm fairly constant, which is nearly 50–150 mM, whereas its concentration in the vacuole fluctuates considerably ranges from 10 to 200 mM or sometimes can extent to 500 mM depending on potash availability. Furthermore, with the increase in plant growth, K percentage (%) in the dry matter (DM) reduces dramatically, and accordingly, it is difficult to diagnose K deficiency as %K in DM without precisely characterization of its optimum concentration at different growth stages of distinct crop species (Johnson and Goulding 1990).

Potassium requirement varies among different crop species according to their capability to acquire it from soil (Table 3.1). Such variation in the uptake of mineral nutrients are accredited to dissimilarity in their root architecture of different plant species, i.e. root branching, density, and rooting depth (Lynch 2019). For instance, wheat and sugar beet usually maintained relatively higher K^+ level in shoot than corn, and thus have a higher K^+ acquisition capacity. Furthermore, wheat can take up higher K^+ from the soil owing to its higher root length to shoot DW ratio, while sugar beet accumulates additional K^+ in the shoot due to three to four times higher K^+ influx than maize and wheat. Furthermore, among all plants, tobacco plant contained the highest K^+ concentration (in leaves and stem) which is nearly 5% on the DW basis.

Table 3.1 Range of sufficient potassium concentrations (upper part) in different plant species

Plant species	K conc. range ($mg\ g^{-1}$ DM)
Cereals: Wheat, maize, rice, oat, barley, rye, etc.	20–58
Dicotyledonous field crops: Sugar beet, potato, cotton, sunflower, peas, soya bean, rape, alfalfa, etc.	17–66
Vegetables: Tomato, cucumber, carrot, lettuce, pepper, spinach, watermelon, onion, etc.	25–60
Fruit trees: Apple, pear, citrus species, etc.	11–30
Forage grasses	25–35
Forest trees: Coniferous trees, broad-leaved trees, etc.	6–15
Miscellaneous crops: Tobacco, tea, vine, etc.	16–50

Source: Bergmann (1993)

3.4 Potassium Deficiency Symptoms in Plants

Potassium is highly mobile which is translocated principally to young growing tissues from older tissues. Therefore, potash deficiency symptoms typically appear first on the lower fully expanded leaves and later on transfer to upper growing leaves as the extent of K deficiency expands. Mengel and Kirkby (2001) further stated that at initial stages of potash deficiency, there is only growth reduction occurred, which is followed by the appearance of chlorosis and necrosis at later stages. On the lower side of leaf, purple spots may appear. The slow growth under K deficiency might be attributed to consequence as K participates in photosynthesis, and protein synthesis processes in the plant (Uchida 2000), which retard under its limited supply.

The appearance of chloroses at the tip of the older fully expanded leaves is the earlier evidence of potash deficiency. In case of cereals, for example, wheat and maize, the appearance of chlorosis on older leaves is the indication of a translocation of K^+ towards younger leaves to sustain the growth up to some extent. As the degree of K deficiency expands, chlorosis start to move from the tip to the edge of leaves and consequently, turn into necrosis (Fig. 3.2). Additionally, plants also display wilting symptoms under severe K deficiency, which are as a result of instable water balance along with a reduced lignification of cell walls. Moreover, K deficient plants are more susceptible to biotic infections, accordingly, reducing the quality of agricultural commodities (see Chaps. 6 and 8 for detail).

In summer crops, the symptoms of K deficiency appear in the start to mid-summer and leaves usually turned pale as like to N deficiency, and upward folding of the edge and curling of leaflets can be observed (Beede et al. 2005). In corn, the lower leaves show chlorotic symptoms with burning edges while midrib of leaf remains green under K deficiency. In additional, its deficiency increases the incidences of various diseases and reduces stalk strength with more lodging, and its deficiency symptoms sometime may confuse with drought or wind scorch. At the early stages of K deficiency, the plant produces many blossom buds, and the fruit quantity and quality can reduce. Furthermore, K deficient fruits and vegetables

Fig. 3.2 Potassium deficiency symptoms in various crop species. Arable crops: (A) wheat, (B) corn, (C) rice, (D) soybean, (E) alfalfa, (F) sweet potato; fruits: (G) apple, (H) strawberry, (I) citrus; vegetables: (J) brinjal, (K) spinach, (L) chilies, (M) onion, (N) tomato, (O) lady finger; cash crops: (P) sugarcane, (Q) tobacco, (R) cotton

deteriorate swiftly when they shipped due to their short shelf life (Ashley et al. 2006).

Potassium deficiency symptoms can be frequently observed in important field crops, vegetables and fruits such as brassica, potatoes, tomatoes, currants, apples, raspberries, gooseberries, clover, and sugar beet. In potato, K deficiency causes the leaves to appear dull and sometimes blue-green in color showing the interveinal chlorosis, which ultimately reduces the tuber size and thus yield. On the upper side, the leaves show a bronzed appearance, while on the underside, they show dark brown spots. In brassica, the leaves become blue-green and show low degree of interveinal chlorosis. The outer edges of the leaves seem to be scorched and leaf texture becomes rough. In alfalfa, the leaflets margins show white or yellow spots. However, K deficiency in alfalfa can be easily confused with the damage from potato leafhopper (Rehm and Schmitt, 2002). In tomatoes, the stems become woody and growth is retarded. The leaves become blue-green, and the interveinal areas sometimes fade to pale gray color.

References

Almodares A, Taheri R, Chung IM, Fathi M (2008) The effect of nitrogen and potassium fertilizers on growth parameters and carbohydrate contents of sweet sorghum cultivars. J Environ Biol 29: 849–852

Arnon DI, Stout PR (1939) The essentiality of certain elements in minute quantity for plants with special reference to copper. Plant Physiol 42:371–385

Ashley MK, Grant M, Grabov A (2006) Plant responses to potassium deficiencies: a role for potassium transport proteins. J Exp Botany 57(2):425–436

Askegaard M, Eriksen J, Johnston AE (2004) Sustainable management of potassium. In: Schjorring P, Elmholt S, Christensen BT (eds) Managing soil quality: challenges in modern agriculture. CABI Publishing, Wallingford, pp 85–102

Beede RH, Brown PH, Kallsen C, Weinbaum SA (2005) Diagnosing and correcting nutrient deficiencies. Pistachio production manual, 4th edn. Ferguson L. University of California, Davis

Chrispeels MJ, Crawford NM, Schroeder JI (1999) Proteins for transport of water and mineral nutrients across the membranes of plant cells. Plant Cell 11:661–676

Hawker JS, Marschner H, Downton WJS (1974) Effects of sodium and potassium on starch synthesis in leaves. Aust J Plant Physiol 1:491–501

Hu W, Liu Y, Loka DA, Zahoor Z, Wang S, Zhou Z (2019) Drought limits pollen tube growth rate by altering carbohydrate metabolism in cotton (*Gossypium hirsutum*) pistils. Plant Sci 286:108–117

Johnson AE, Goulding KW (1990). The use of plant and soil analysis to predict the potassium supplying capacity of soil. In: Development of K-fertilizer recommendations: proceedings 22nd Colloquium of the International Potash Institute, p 177–204

Lynch JP (2019) Root phenotypes for improved nutrient capture: an underexploited opportunity for global agriculture. New Phytol 223:548–564

Marschner P (2012) Marschner's mineral nutrition of higher plants, vol 89. Academic Press, Oxford

Mengel K (2007) Potassium. In: Barker AV, Pilbeam DJ (eds) Handbook of plant nutrition. Taylor & Francis, Boca Ratan, FL

Mengel K, Kirkby EA (2001) Principles of plant nutrition, 5th edn. Kluwer Academic Publishers, Dordrecht

Miller C (1993) Potassium selectivity in proteins: oxygen cage or pi in the face. Science 261:1692–1693

Oosterhuis DM, Loka DA, Raper TB (2014) Potassium and stress alleviation: physiological functions and management of cotton. J Plant Nutr Soil Sci 176:1–13

Pettigrew WT (2008) Potassium influences on yield and quality production for maize, wheat, soybean and cotton. Physiol Plantarum 133:670–681

Raddatz N, de los Ríos LM, Lindahl M, Quintero FJ, Pardo JM (2020) Coordinated transport of nitrate, potassium, and sodium. Front Plant Sci 11:247

Rehm G, Schmitt M (2002) Potassium for crop production. http://www.extension.umn.edu/agriculture/nutrient management/potassium/potassium-for-crop-production/.

Schubert S (2006) Pflanzenernahrung. Verlag Eugen Ulmer, Stuttgart

Taiz L, Zeiger E (2010) Plant physiology, 5th edn. Sinauer Associates, Sunderland, MA

Thomas TC, Thomas AC (2009) Vital role of potassium in the osmotic mechanism of stomata aperture modulation and its link with potassium deficiency. Plant Signal Behav 4(3):240–243

Uchida R (2000) Essential nutrients for plant growth: nutrient functions and deficiency symptoms. In: Plant nutrient management in Hawaii's soils. College of Tropical Agriculture and Human Resources, University of Hawaii, Manoapp, pp 31–55

Van Brunt JM, Sultenfuss JH (1998) Better crops with plant food. Potassium: functions of potassium 82(3):4–5

Wang M, Zheng Q, Shen Q, Guo S (2013) The critical role of potassium in plant stress response. Int J Mol Sci 14:7370–7390

Yamaguchi T, Blumwald E (2005) Developing salt-tolerant crop plants: challenges and opportunities. Trends Plant Sci 10:615–620

Yang J, Hu W, Zhao W, Chen B, Wang Y, Zhou Z, Meng Y (2016) Fruiting branch K^+ level affects cotton fiber elongation through osmoregulation. Front Plant Sci 7:13

Phytoavailability of Potassium

4

4.1 Acquisition and Uptake by Plants

Potassium is acquired by plants as cation from soil solution through plasma membranes of cortical and epidermal root cells, and it is transported towards the xylem through symplastic movement. Indeed, the K delivery towards the surface of the roots and K uptake kinetics by the root cells principally impact the acquisition of K by crop plants (White and Karley 2010). Both transpiration-driven mass flow and the local diffusion through the rhizosphere are responsible for the K delivery to the root surface (Tinker and Nye 2000). The K present in the rhizosphere entered the plasma membranes of the root cells through K transporters (Britto and Kronzucker 2008; Karley and White 2009; White and Karley 2010). The relationship between the K uptake and its concentration within the rhizosphere generally follows the sum of a linear and hyperbolic function, which can be attributed to various K transporter activities, having various kinetic properties (Britto and Kronzucker 2008; White and Karley 2010).

Various factors such as release of root exudates, root proliferation, the rate of K uptake in root plasma membrane, and transpiration rate manipulates the acquisition and uptake of K by plants. For instance, root exudates including the organic acids solubilize the non-exchangeable K, thus enhancing the concentration of K within the soil. The enhanced root proliferation within the soil increase the surface area for K uptake, with simultaneous decreases in the distance required for water flow and diffusion of K. Likewise, the rate of uptake of K across the plasma membranes of roots impacts the concentration of K in rhizospheric solution and the rate of diffusion of K^+ towards the surface of the roots. The rate of transpiration governs the mass flow of soil solution (loaded with K^+) towards the roots (Tinker and Nye 2000; Rengel and Damon 2008; White and Karley 2010).

The crop growth and yield follow the law of diminishing returns, with improvement in K phytoavailability. If the ammonical N is absent in the rhizosphere, and the supply of K to the roots is in equilibrium with the minimal K demand by the plant,

A. Wakeel, M. Ishfaq, *Potash Use and Dynamics in Agriculture*,
https://doi.org/10.1007/978-981-16-6883-8_4

the roots of plants will acquire sufficient K (present in micromolar concentrations) from soil solutions to continue their growth cascades (White and Karley 2010). There exists a specific time course for the uptake and internal redistribution of K during the life cycle of the plants. The K reserves within seed support the initial seedling growth; the following developmental stages being dependent on the uptake of K via roots (White and Veneklaas 2012). Thus, the K concentration within plant tissues is directly related to the roots ability to obtain K from the soil solution. The patterns of accumulation of K differ among the plant species and even among the various cultivars of the same crop. Field crops such as rice, maize, canola, and wheat contain maximum K at the anthesis stage which is translocated from senescing leaves towards the seeds. However, few field crops such as lupine, sunflower, soybean, and sorghum continue to obtain the significant amount of K from the soil solution during the process of maturation of seeds (Rose et al. 2007, 2008; Fageria et al. 2011).

Various morphological, physiological, and biochemical responses occur within plants to enhance the acquisition of K from soil if its acquisition from plant roots is insufficient to cope with plant growth. For instance, the increase in number, activity, and affinity for root plasma membrane K transporters has been observed to fulfill the K demands of the plants (White and Karley 2010). However, in the soil having high K availability, enhancement in the number/activity of K transporters will be beneficial; whereas, in the soils with low K concentration in soil solution, enhancement in the affinity K transporters will be beneficial to improve K acquisition. The physiological strategies during low K supply includes the vacuolar K replacement with some alternative osmotica, translocation of K towards developing tissues from the mature tissues and decreasing the growth in order to maintain adequate tissue K concentration to continue the cell functions. At low K levels, plants enhance the concentration of amide nitrogen, sugars (e.g. sucrose), polyamines (e.g. agmatine and putrescine), and amino acids, e.g. tyrosine, arginine, and lysine, and readily available cations, i.e. sodium, calcium, magnesium, and may be some heavy metals (Amtmann et al. 2006; White and Karley 2010; Wakeel et al. 2010, 2011), which further decrease the tissue K requirements.

4.2 Transport and Accumulation: Underlying Players

When K reached the root surface, the kinetics for uptake and transport of K^+ within the plant is achieved through two mechanisms. These systems are known as high-affinity transport system (HATS) and low-affinity transport system (LATS; channel mediated) (Rodríguez-Navarro 2000; Britto and Kronzucker 2008). The high-affinity K^+ uptake is controlled by either the KT/HAK/KUP family, i.e. HvHAK1 in barely, or the HKT family such as TaHKT1 in wheat (Schachtman and Schroeder 1994; Santa-Maria et al. 2000). Commonly, the LATS work at external K^+ concentration more than 0.3 mM (Wang and Wu 2013), but not exceeds millimolar range. The LATS activity is, generally, insensitive to the external K input (Chérel et al. 2014). However, upon the stop of exogenous K supply, HATS is rapidly upregulated. Numerous research studies confirmed that these two systems are carrying out their activities in different plant species and are responsible for uptake and transport of K^+ in the plants (Fig. 4.1).

In HATS, the K enters the cell through symport with H^+ with a proposed "1:1 stoichiometry" in an energy-dependent process (Maathuis et al. 1997), with the involvement of the transmembrane proton motive force. The H^+ and the electrical gradient underlying this transmembrane proton motive force are preserved by membrane-bound ATP-hydrolyzing transporters, which pump H^+ towards the external medium from cytosol (Palmgren 2001; Pardo et al. 2006). Several factors such as gene transcription and binding of substrate with carrier molecule regulate the HATS (Gierth et al. 2005). Whereas in LATS, K^+ is transported through selective and non-selective ion channels, through the formation of pores, and that cross the membrane bilayer. Thermodynamically, the activities of these channels rely on the electrochemical potential gradients for K^+ to drive the transport and is regulated by many factors such as pH, membrane voltage, cyclic nucleotides, reactive oxygen species, carbon dioxide dependent light activation, and K^+ itself (Talke et al. 2003; Cakmak 2005; Liu et al. 2006; Johansson et al. 2006).

The transmembrane electrical potential can drive the entire K^+ transport at high and mid-range of external concentrations, but acquirement of K^+ from K^+-deprived soil needs an extra source of energy, which can be delivered through the coupled transport (via carrier-type transporter) or through a hypothetical K^+-motive ATPase. Some findings have suggested the role of K channels in high-affinity K^+ uptake (Sentenac et al. 1992). For example, in akt1-1 arabidopsis mutant, the disruption of AKT1 K channels retards the growth of plants at low external K^+, explicitly 10 μM (Hirsch et al. 1998). The transport properties of KAT1 and AKT1 channels indicate that these channels are expressed in the heterologous system in which they probably take up K^+ in those tissues of the plant which express these channels; primarily in the case of KAT1 guard cells and for AKT1 root cells (Bertl et al. 1997; Gaymard et al. 1998). Furthermore, the K^+ transport capacity of this heterologous system depends on the concentrations of H^+ and K^+ in the medium and is mediated by the increased concentration of H^+ and Na^+ (Spalding et al. 1999).

Potassium uptake in plants may be strengthened through the proton-electrochemical gradient, inwardly directed, and is utilized by some KT/KUP/HAK transporters, which serve as co-transporter for H^+ and K^+ (Rodríguez-Navarro 2000). According to Palmgren (2001), the proton pumps in the plasma membrane provide motive force in the form of H^+ electrical gradient across the membrane. The role of electrochemical K^+ gradient in K supply is very vital. Even at the very low concentration of external K, the low electrochemical potential gradient of cytosol improves the K entry into cells via K^+ channels (voltage gated). The cytosolic positive charge increases under the entry of K^+ and decrease negative charge of the cytosolic, finally equilibrium condition is attained with the outer medium. Mengel and Kirkby (2001) revealed that there is no driving force for K^+ uptake under these equilibrium conditions. For example, in wheat, barley, and rice, the transcripts encoding the HKT1 over expressed when K was deficient, but high-affinity uptake of Na^+ was increased (Garciadeblás et al. 2003; Haro et al. 2010).

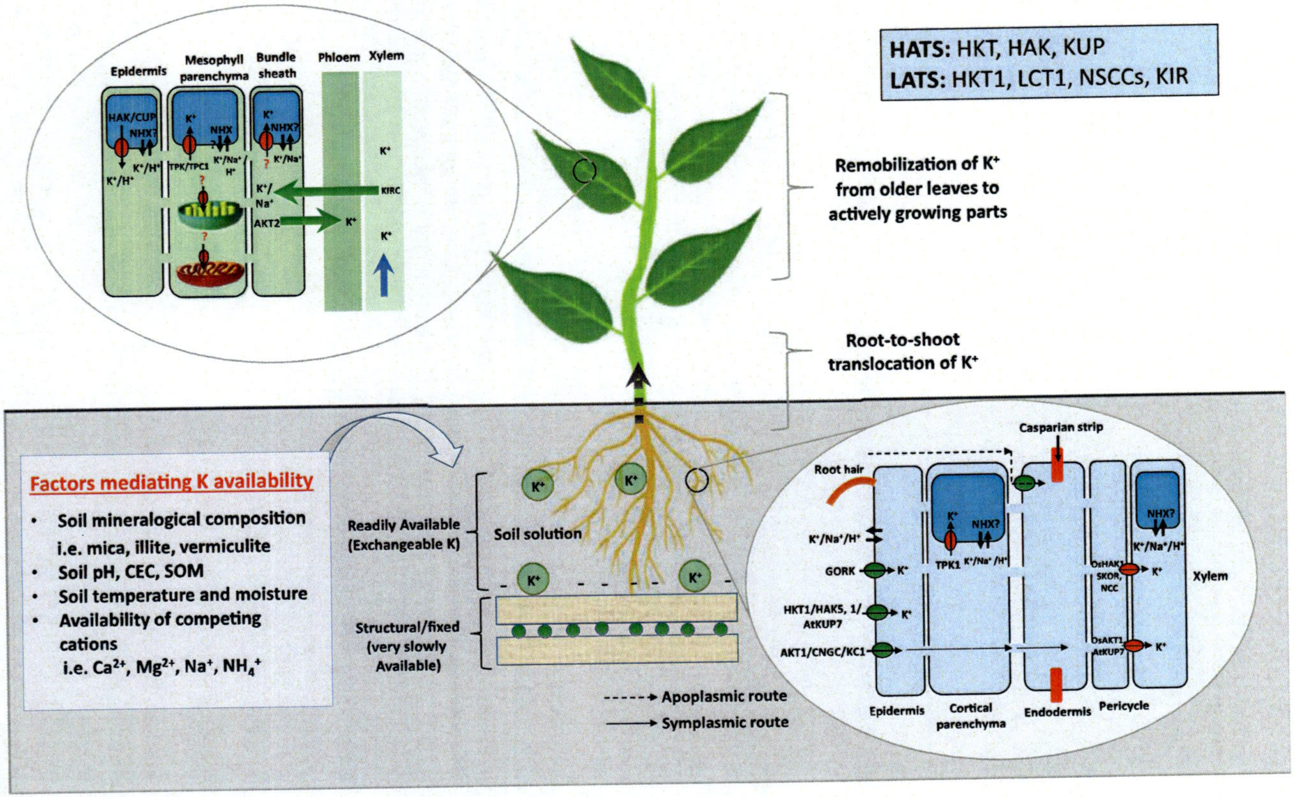

Fig. 4.1. Potassium uptake, translocation, and homeostasis in plants. Plant take up K in cationic form from the soil solution that is mediated by different soil environmental stimuli, for instance, soil mineralogical composition, temperature, moisture, or the presence of competing cations (i.e. Ca^{2+}, Mg^{+2}, Na^{+}, NH_4^{+}).

Once the K^+ reached the root surface, plant take up by different transporters/exchanger, followed by enter into root epidermis, cortex and then to enter the xylem, it must pass through the casparian strip. Later on, it transfers to shoot and leaves through xylem and store in vacuole via different transporters/exchanger (HATS: high-affinity transport system; LATS: low-affinity transport system)

The membrane transporters in plants responsible for K flux has been divided into three channels and three transporters based on phylogenetic analysis (Lebaudy et al. 2007; Alemán et al. 2011). A K channel (AKT1), expressed in *Arabidopsis thaliana*, triggers the absorption of K under the wide range of external K concentration (Hirsch et al. 1998; Spalding et al. 1999). Some studies have reported that many transporters including the members of K transporters (KT), K uptake permease (KUP), and high-affinity K transporters family are low-affinity K transporters (Senn et al. 2001). Among the three families of K transporters, high-affinity K/Na transporter (HKT), cation proton antiporters (CPA), and KT/HAK/KUP are well studied. The CPA serve as a $K^+(Na^+)/H^+$ antiporter, while the HKT may cotransport K and Na or transport Na only. The KT/HAK/KUP are the H^+-coupled K^+ symporters, actually (Mäser et al. 2001; Lebaudy et al. 2007).

Some recent transcriptomic findings have reported various K transporters which triggers the influx of K from rhizosphere to cells, and efflux of K towards xylem form stellar cells (Britto and Kronzucker 2008; Karley and White 2009; White and Karley 2010). By utilizing the voltage and pH gradient by the plasma membrane H^+-ATPases, H^+/K^+ symporters encoded by the members of CHX and KT/HAK/HUP gene families, can catalyze the influx of K to the root cells against its electrochemical gradient when the concentration of rhizosphere K is more than 100 nM. At the low concentration of rhizospheric K, the orthologs of the H^+/K^+ symporter, for instance, AtHAK5 of *Arabidopsis thaliana* (induced by K deficiency) has a significant role in K nutrition to plants (White and Karley 2010; Alemán et al. 2011). When the concentration of rhizospheric K exceeds 1 mM, multiple K-permeable channels are in action by using the voltage gradient across the plasma membrane, catalyze the influx of K to root (Karley and White 2009; White and Karley 2010). The orthologs of the inward-rectifying K channels, which are formed by AtKH1 and AtAKT1 in *Arabidopsis thaliana* play vital role in K nutrition to plants (Pyo et al. 2010; Rubio et al. 2010; Alemán et al. 2011).

In plants, the KT/HAK/KUP are recognized as the largest family of K transporters and these transporters might be coordinatively involved in K^+ uptake/transport in different plant species (Mäser et al. 2001; Bañuelos et al. 2002; Gupta et al. 2008; Xie et al. 2020). The sequence alignment shows that the genes of this family are homologous to each other up to a certain extent. The KT/HAK/KUP family is further separated into four major clusters, i.e. clusters I, II, III, and IV (Rubio et al. 2000; Gupta et al. 2008), where clusters I and II further differentiated into groups A and B. The genes of the clusters I and II are usually found in all plants. The cluster III is comprised of genes from both arabidopsis and rice, while the cluster IV is comprised of only 4 rice genes (Grabov 2007; Gupta et al. 2008). These transporters also help plants to within various abiotic stresses. For example, in tobacco (*Nicotiana tabacum*), the expression of OsHAK5 improved the tolerance against salinity through the enhanced accumulation of K without impacting the Na contents (Horie et al. 2011).

Potassium selective channels are considered as multimers (composed of a-subunits) and belongs to three major families, viz. TPK (tandem-pore), the shaker, and the KIR-K inward-rectifying (Lebaudy et al. 2007). For instance, in grape vine shaker K^+ channel VvK3.1 supports its function in massive K fluxes, which is necessary for berry K loading (Nieves-Cordones et al. 2019). Besides these K-specific channels families, K^+ currents can also be conducted through the non-selective cation channels (Demidchik and Maathuis 2007). These channels include the cyclic-nucleotide-gated channels (CNGs) and the Shaker-like channels. In arabidopsis, 20 CNGs have been identified (Talke et al. 2003) and among these 20 CNGs, the putative K^+ channel (i.e. AtCNGC10) has recently been found to have great impact on growth and accumulation of starch (Borsics et al. 2007). Furthermore, two electroneutral CPA, viz. CPA1 and CPA2 have been found to regulate the non-channel-mediated transport of K (Gierth and Mäser 2007).

4.3 Factors Affecting Potassium Availability to Plants

Potassium availability to plants is mediated by a wide range of intrinsic and extrinsic factors such as soil type, fertilizer, and other management practices. For instance, adequate soil moisture supply is necessary to facilitate diffusion of K (which usually accounts for >75% of K movement) to plant roots for uptake. Mass flow, which also accounts for some soil K transport, also requires sufficient water in the soil. Therefore, soil moisture deficits can limit soil K transport as well as uptake into the plant, thereby causing K deficiency. Potash uptake also depends on plant factors, including genetics and developmental stage (Rengel and Damon 2008). Competition for photoassimilates between developing fruits and vegetative organs during reproductive growth stages can limit root growth/activity and K uptake. Under such conditions, increasing soil K fertilization may not be enough to alleviate this developmentally induced deficiency partly because of reduced root growth/activity during reproductive development and also because of competition from other cations for binding sites on roots (Marschner 2012).

4.3.1 Soil Factors

The soil factors affecting the K availability include the history of soil formation, the cation exchange capacity of the soil, the amount of available K within soil, non-exchangeable or slowly available K, soil K fixation capacity, the quantity of K in subsoil and the consistency/density of subsoil layers, soil temperature, soil moisture, soil aeration, and soil tilth. The history of soil formation is very crucial that impacts the K availability to plants. This includes the material from which soil is formed, clay mineral types, the climate and vegetation under which soil is formed, soil drainage and topography. If the soil cation capacity will be high, the supplying power and the storage capacity for K will be high and vice versa. Similarly, if the total amount of available K will be high, the availability to plants will be high. As, the non-exchangeable/slowly available K renews the supply of exchangeable K in the soil, thus high amount of non-exchangeable K in soil ensures more renewal of exchangeable K.

Soil texture has a strong influence on K availability. For example, clay soils may have high K-fixing capacities and thus can show little response to soil-applied K fertilizers because much of the available K quickly binds to clays (Brady and Weil 1999). Such K retention can support to reduce leaching losses and be beneficial in the long term as storage reservoirs of K for subsequent crops. However, soil enriched in a clay type soil fraction may fix huge quantities of K applied through fertilizer which reduces K availability to the crop so is the plant response to K fertilization. Likewise, in some areas, K availability from the subsoil is high if the roots penetrate deeper and is low if the roots are not able to penetrate due to soil compaction. In contrast, sandy soils usually have a low K-supplying power because of low cation exchange capacity. Additionally, calcareous soils tend to have high concentrations of calcium ions (Ca^{2+}) that dominate clay surfaces and other exchange sites. The antagonistic tendencies of cationic nutrients, particularly Ca^{2+} and magnesium (Mg^{2+}) tend to limit K^+ uptake by competing for binding sites on root surfaces. Consequently, crops grown on highly calcareous soils can show K deficiency symptoms even though the soil test may report enough K (Havlin et al. 1999).

The various functions of the plant body, the activity of roots, and different physiological processes within plant increases as the temperature of the soil increases. Improvement in physiological activities thus improves the uptake of K. The optimum soil temperature for the K uptake is 60–80°F, and its uptake is reduced at low soil temperatures (Rehm and Schmitt 2002). Thus, row placemen of K will be beneficial at low soil temperatures, for minimum till and early planted crops. As the soil moisture level increases, the availability of K to the plants also increases. Indeed, improvement in soil moisture improves the movement of K towards the plant root through mass flow and its diffusion to roots, thus enhances its availability to plants (Rehm and Schmitt 2002). Furthermore, soil aeration is very important for the respiration of roots and uptake of K. In brief, when the soil moisture moves towards the saturation, the activity of roots and the subsequent uptake of K decreases owing to a low level of oxygen in saturated soil (Rehm and Schmitt 2002). Besides, tillage in wet soils may cause soil compaction thus affecting the soil aeration, oxygen level, and K availability.

4.3.2 Fertilizer and Crop Management Practices

Fertilizer and crop management practices affect the K availability and uptake by plants. When soil adequate quantities of available K^+, the addition of phosphorous and nitrogen (N) will enhance the uptake of K. Usually, the K^+ uptake by crops is equal to N uptake or even greater than N in many cases. Application of K^+ through crop residues, manure and artificial fertilizer enhances the K^+ availability to plants as K^+ from these sources is easily available to plants. The placement of K^+ fertilizer is very important regarding its availability to plants (See Chap. 8 for detail). In brief, K applied through broadcast low-down method is easily available to the plants rather than the surface applied disked-in method. Application of K at deeper depth and drip irrigation also facilities the downward movement of K, thus enhancing its availability to roots. Furthermore, in fine textured soil, the application of gypsum is very beneficial to move the K down to make it available for plant uptake by roots. Soil drainage improves the soil aeration and thus enhances the K availability to plants. Keeping weeds and insect pest below threshold level also reduces the competition for water and nutrients and thus improves the K availability to plants.

4.3.3 Abiotic Stresses

Abiotic stresses such as drought, water logging, and salinity reduce the K^+ availability to plants. According to Hu and Schmidhalter (2005), drought and salinity stresses reduce the K^+ uptake in plants. Thus, maintaining the optimum K^+ status within plants is necessary to cope the drought stress, as the plant under drought stress has a huge requirement of internal K^+ (Cakmak 2005). The reduction in K uptake due to drought stress might be attributed to reduce K mobility in soil, decreased transpiration rate, and impaired activity of transporters of the root plasma membrane (Hu and Schmidhalter 2005). Similarly, K^+ uptake and transport is reduced under saline condition (Tavakkoli et al. 2011), owing to the decrease in water uptake due to the difference in water potential between the soil and the plant body. Under waterlogged conditions, there is deficiency of oxygen which reduces aeration and eventually K^+ uptake (Board 2008). Indeed, water logging causes the impairment in the ion transport process, taking place through the ion transport and channel, thus decreasing the uptake of essential cations including K^+ (Colmer and Greenway 2011).

References

Alemán F, Nieves-Cordones M, Martínez V, Rubio F (2011) Root K^+ acquisition in plants: the Arabidopsis thaliana model. Plant Cell Physiol 52:1603–1612

Amtmann A, Hammond JP, Armengaud P, White PJ (2006) Nutrient sensing and signalling in plants: potassium and phosphorus. Adv Bot Res 43:209–257

Bañuelos MA, Garciadeblas B, Cubero B, Rodríguez-Navarro A (2002) Inventory and functional characterization of the HAK potassium transporters of rice. Plant Physiol 130:784–795

Bertl A, Reid JD, Sentenac H, Slayman C (1997) Functional comparison of plant inward rectifier channels expressed in yeast. J Exp Bot 48:405–413

Board JE (2008) Waterlogging effects on plant nutrient concentrations in soybean. J Plant Nutr 31(5):828–838

Borsics T, Webb D, Andeme-Ondzighi C, Staehelin LA, Christopher DA (2007) The cyclic nucleotide-gated calmodulin-binding channel AtCNGC10 localizes to the plasma membrane and influences numerous growth responses and starch accumulation in Arabidopsis thaliana. Planta 225:563–573

Brady NC, Weil RR (1999) The nature and properties of soils, 9th edn. Macmillan Publishing Company, New York, p 750

Britto DT, Kronzucker HJ (2008) Cellular mechanisms of potassium transport in plants. Physiol Plantarum 133:637–650

Cakmak I (2005) The role of potassium in alleviating detrimental effects of abiotic stresses in plants. J Plant Nutr Soil Sci 168:521–530

Chérel I, Lefoulon C, Boeglin M, Sentenac H (2014) Molecular mechanisms involved in plant adaptation to low K^+ availability. J Exp Botany 65:833–848
Colmer TD, Greenway H (2011) Ion transport in seminal and adventitious roots of cereals during O_2 deficiency. J Exp Botany 62(1):39–57
Demidchik V, Maathuis JM (2007) Physiological roles of nonselective cation channels in plants: from salt stress to signaling and development. New Phytol 175:387–404
Fageria NK, Baligar VC, Jones CA (2011) Growth and mineral nutrition of field crops. CRC Press, Boca Raton, FL
Garciadeblás B, Senn ME, Bañuelos MA, Rodríguez-Navarro A (2003) Sodium transport and HKT transporters: the rice model. Plant J 34(6):788–801
Gaymard F, Pilot G, Lacombe B, Bouchez D, Bruneau D, Boucherez J, Michaux-Ferriere N, Thibaud JB, Sentenac H (1998) Identification and disruption of a plant shaker-like outward channel involved in K^+ release into the xylem sap. Cell 94:647–655
Gierth M, Mäser P (2007) Potassium transporters in plants –involvement in K1 acquisition, redistribution and homeostasis. FEBS Lett 581:2348–2356
Gierth M, Mäser P, Schroeder JI (2005) The potassium transporter AtHAK5 functions in K1 deprivation-induced high-affinity K1 uptake and AKT1 K1 channel contribution to K1 uptake kinetics in Arabidopsis roots. Plant Physiol 137:1105–1111
Grabov A (2007) Plant KT/KUP/HAK potassium transporters: single family multiple functions. Ann Bot (Lond) 99:1035–1041
Gupta M, Qiu X, Wang L, Xie W, Zhang C, Xiong L, Lian X, Zhang Q (2008) KT/HAK/KUP potassium transporters gene family and their wholelife cycle expression profile in rice (*Oryza sativa*). Mol Gen Genomics 280:437–452
Haro R, Bañuelos MA, Rodríguez-Navarro A (2010) High-affinity sodium uptake in land plants. Plant Cell Physiol 51(1):68–79
Havlin JL, Beaton JD, Tisdale SL, Nelson WL (1999) Soil fertility and fertilizers, 6th edn. Prentice Hall, Upper Saddle River, NJ, p 499
Hirsch RE, Lewis BD, Spalding EP, Sussman MR (1998) A role for the AKT1 potassium channel in plant nutrition. Science 280:918–921
Horie T, Sugawara M, Okada T, Taira K, Kaothien-Nakayama P, Katsuhara M, Shinmyo A, Nakayama H (2011) Rice sodium-insensitive potassium transporter, OsHAK5, confers increased salt tolerance in tobacco BY2 cells. J Biosci Bioenergy 111:346–356
Hu Y, Schmidhalter U (2005) Drought and salinity: a comparison of their effects on mineral nutrition of plants. J Plant Nutr Soil Sci 168(4):541–549
Johansson I, Wulfetange K, Poree F, Michard E, Gajdanowiczm P, Lacombe B, Sentenac H, Thibaud JB, Mueller-Roeber B, Blatt MR, Dreyer I (2006) External K1 modulates the activity of the Arabidopsis potassium channel SKOR via an unusual mechanism. Plant J 46:269–281
Karley AJ, White PJ (2009) Moving cationic minerals to edible tissues: potassium, magnesium, calcium. Curr Opin Plant Biol 12:291–298
Lebaudy A, Very AA, Sentenac H (2007) K1 channel activity in plants: genes, regulations and functions. FEBS Lett 581:2357–2366
Liu K, Li LG, Luan S (2006) Intracellular K1 sensing of SKOR, a shaker-type K1 channel from Arabidopsis. Plant J 46:260–268
Maathuis FJM, Sanders D, Gradmann D (1997) Kinetics of high-affinity K1 uptake in plants, derived from K1-induced changes in current-voltage relationships – a modelling approach to the analysis of carrier-mediated transport. Planta 203:229–236
Marschner P (2012) Marschner's mineral nutrition of higher plants, vol 89. Academic Press, Oxford, UK
Mäser P, Thomine S, Schroeder JI, Ward JM, Hirschi K, Sze H (2001) Phylogenetic relationships within cation transporter families of Arabidopsis. Plant Physiol 126:1646–1667
Mengel K, Kirkby EA (2001) Principles of plant nutrition, 5th edn. Kluwer Academic Publishers, Dordrecht, p 849
Nieves-Cordones M, Andrianteranagna M, Cuellar T, Cherel I, Gibrat R, Boeglin M, Moreau B, Paris N, Verdeil JL, Zimmermann SD, Gaillard I (2019) Characterization of the grapevine

shaker K^+ channel VvK3.1 supports its function in massive potassium fluxes necessary for berry potassium loading and pulvinus-actuated leaf movements. New Phytol 222:286–300
Palmgren MG (2001) Plant plasma membrane H1-ATPases: powerhouses for nutrient uptake. Annu Rev Plant Physiol Plant Mol Biol 52:817–845
Pardo JM, Cubero B, Leidi EO, Quintero FJ (2006) Alkali cation exchangers: roles in cellular homeostasis and stress tolerance. J Exp Botany 57:1181–1199
Pyo YJ, Gierth M, Schroeder JI, Cho MH (2010) High-affinity K^+ transport in Arabidopsis: AtHAK5 and AKT1 are vital for seedling establishment and post germination growth under low potassium conditions. Plant Physiol 153:863–875
Rehm G, Schmitt M 2002. Potassium for crop production. http://www.extension.umn.edu/agriculture/nutrient management/potassium/potassium-for-crop-production/
Rengel Z, Damon PM (2008) Crops and genotypes differ in efficiency of potassium uptake and use. Physiol Plantarum 133(4):624–636
Rodríguez-Navarro A (2000) Potassium transport in fungi and plants. Biochim Biophys Acta-Biomembr 1469:1–30
Rose TJ, Rengel Z, Ma Q, Bowden JW (2007) Differential accumulation patterns of phosphorus and potassium by canola cultivars compared to wheat. J Plant Nutr Soil Sci 170:404–411
Rose TJ, Rengel Z, Ma Q, Bowden JW (2008) Post-flowering supply of P, but not K, is required for maximum canola seed yields. Eur J Agronomy 28:371–379
Rubio F, Santa-Maria GE, Rodriguez-Navarro A (2000) Cloning of Arabidopsis and barley cDNAs encoding HAK potassium transporters in root and shoot cells. Physiol Plantarum 109:34–43
Rubio F, Alemán F, Nieves-Cordones M, Martínez V (2010) Studies on Arabidopsis athak5, atakt1 double mutants disclose the range of concentrations at which AtHAK5, AtAKT1 and unknown systems mediate K^+ uptake. Physiol Plantarum 139:220–228
Santa-Maria GE, Danna CH, Czibener C (2000) High-affinity potassium transport in barley roots. Ammonium-sensitive and -insensitive pathways. Plant Physiol 123:297–306
Schachtman DP, Schroeder JI (1994) Structure and transport mechanism of a high-affinity potassium uptake transporter from higher-plants. Nature 370:655–658
Senn ME, Rubio F, Bañuelos MA, Rodriguez-Navarro A (2001) Comparative functional features of plant potassium HvHAK1 and HvHAK2 transporters. J Biol Chem 276:44563–44569
Sentenac H, Bonneaud N, Minet M, Lacroute F, Salmon JM (1992) Cloning and expression in yeast of a plant potassium ion transport system. Science 256:663–665
Spalding EP, Hirsch RE, Lewis DR, Qi Z, Sussman MR, Lewis BD (1999) Potassium uptake supporting plant growth in the absence of AKT1 channel activity – inhibition by ammonium and stimulation by sodium. J Gen Physiol 113:909–918
Talke IN, Blaudez D, Maathuis FJM, Sanders D (2003) CNGCs: prime targets of plant cyclic nucleotide signalling? Trends Plant Sci 8:286–293
Tavakkoli E, Fatehi F, Coventry S, Rengasamy P, McDonald GK (2011) Additive effects of Na^+ and cl–ions on barley growth under salinity stress. J Exp Botany 62(6):2189–2203
Tinker PB, Nye PH (2000) Solute movement in the Rhizosphere. Oxford University Press, New York
Wakeel A, Steffens D, Schubert S (2010) Potassium substitution by sodium in sugar beet (*Beta vulgaris*) nutrition on K-fixing soils. J Plant Nutr Soil Sci 173:127–134
Wakeel A, Farooq M, Qadir M, Schubert S (2011) Potassium substitution by sodium in plants. Crit Rev Plant Sci 30:401–413
Wang Y, Wu WH (2013) Potassium transport and signaling in higher plants. Annu Rev Plant Biol 64:451–476
White PJ, Karley AJ (2010) Potassium. In: Hell R, Mendel RR (eds) Plant cell monographs 17. Cell biology of metals and nutrients. Springer, Berlin, pp 199–224
White PJ, Veneklaas EJ (2012) Nature and nurture: the importance of seed phosphorus. Plant Soil 357:1–8
Xie Q, Ma L, Tan P, Deng W, Huang C, Liu D, Lin W, Su Y (2020) Multiple high-affinity K^+ transporters and ABC transporters involved in K^+ uptake/transport in the potassium-hyperaccumulator plant Phytolacca acinosa Roxb. Plants (Basel, Switzerland) 9:470

5 Potassium Evaluation in Soils

Potassium evaluation in soils resembles to other mineral nutrients and principally any approach can be followed depending on time and available resources. In addition to crop response to potash fertilization, there are three distinct evaluation methods, namely visual observations of plants, soil testing, and plant–tissue analysis which can be base line to characterize the K nutritional status of plants.

5.1 Visual Observations of Plants

Precise observations of the physical appearance of plants can support the direct indication to predict their nutritional status. Under the deficiency of a nutrient, metabolic disruptions provide associations between the role of that nutrient and the appearance of a certain visual abnormalities in plants. Therefore, distinct disorder symptoms can specify an evidence to categorize nutritional status of the specific nutrient. Furthermore, deficiency symptoms of any nutrient can appear at any growth stage of the plant according to the extent of the deficiency. However, the earlier diagnosis of nutritional deficiencies can allow time for corrective measures to take place. Nevertheless, precise experimental work and observations are required to distinguish deficiency symptoms of the nutrient. Furthermore, to characterize the deficiency or toxicity symptoms of any nutrient, it is essential to be familiar with the specific role and mobility of a mineral nutrient within plant.

The deficiency symptoms of mobile nutrients usually appear first on older matured leaves because they translocate it from older leaves towards the actively growing parts of plant; in contrast, immobile nutrients usually show deficiency symptoms on younger upper leaves. As K is literally mobile nutrient in plants, thus, its deficiency symptoms first appear on older fully expanded mature leaves. The distinct deficiency symptoms of K can be leaf chlorosis which first appear at the margins of older fully expanded leaves that lead to necrotic spots (Barnes et al. 2012), falling of lower leaves, stunted plant growth with small branches, rusty brown

A. Wakeel, M. Ishfaq, *Potash Use and Dynamics in Agriculture*,
https://doi.org/10.1007/978-981-16-6883-8_5

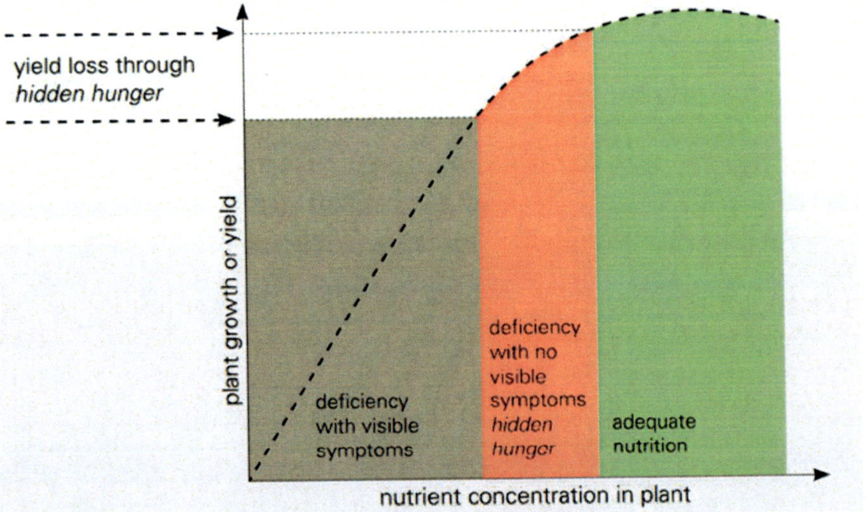

Fig. 5.1 Illustration of the effect of K deficiency on growth or yield, with and without visible symptoms (Adapted from IPI booklet "*Potassium a Nutrient Essential for Life*")

spots on the older fully expanded leaves, stem lodging, ultimately result in poor yield and quality of agricultural produce (see Fig. 3.2 for visual K deficiency symptoms in different crop species). Furthermore, severe K deficiency even can cause death of plant tissue. As potash is also involve in the translocation of photo assimilates, hence, its deficiency can cause impaired carbohydrate allocation between source and sink organs. Finally, results in the reduction root growth and other reproductive parts of plant, for instance, grains, fruits.

Potassium deficiency symptoms are frequently experienced in early growth stages and its effect usually depends on crop species. For instance, tomato exhibit uneven fruit ripening, while leaves of cotton turn reddish brown and then black with the progression of deficiency, and finally fell, and yield and quality of forage crops are negatively effect. In maize crop, K deficiency symptoms usually appear as yellow and brown leaf margins, nevertheless, its deficiency symptoms are difficult to characterize at maturity or harvest stage. At harvest stage, maize shows few visual indications of potash deficiency, for instance, the broken stalk, unfilled cob tips, and smaller cobs. In sugarcane crop, reddish leaf margins or veins, stunted growth, lodging, reduced sucrose contents (sugar recovery) are owing to K deficiency that are commonly observed in many sugarcane fields in Pakistan. However, crops can be deficient in K without exhibiting any visible symptoms; this effect is commonly referred to as "hidden hunger," that can considerably reduce crop yield and quality without any apparent deficiency symptoms (Fig. 5.1).

Although characterization of visual symptoms in plants is a quick and inexpensive way to detect and characterize deficiency or toxicity of the nutrient in field. However, visual observation can be diagnostic approach, but not an authenticated approach and can be restricted by several factors, such as pseudo deficiency and hidden hunger. Furthermore, experience is required to characterize the distinct deficiency symptoms of a specific nutrient, they can be misdiagnosed with closely resemble deficiency symptoms of other nutrients. In addition, symptoms that look like nutrient deficiency can appear on plants by means of other biotic or abiotic factors which are not actually deficiency of any nutrient, possibly, drought or waterlogging, disease or insect infestations, mechanical/herbicide/wind or any other damage, salinity or elemental toxicities. Therefore, plant leaf or tissue analysis, soil analysis, and plant response to specific nutrient are additional approaches to characterize the actual deficiency or toxicity of a distinct mineral nutrient.

Furthermore, if deficiency symptoms of specific nutrient have appeared or irreversible damage has occurred, the quantity or quality of agricultural commodities are reduced or diminished at later stage, it will be difficult to attain a desirable yield with application of required fertilizer. Indeed, under K deficiency plants only accumulate sucrose in their source leaves, that distinguish it from other nutrient deficiency, for instance, under N or P deficiency plant only accumulate starch (Cakmak and Kirkby 2008; Verbruggen and Hermans 2013), while under Mg deficiency plants accumulate sucrose as well as starch in their source leaves (Tanoi and Kobayashi 2015). The accumulation of sucrose in the source leaves is an earlier response of plant to K deficiency rather than the appearance of deficiency symptoms. Additionally, plants can be verified for availability of nutrients by quantitative or analytic analysis (distinct nutrient along with total concentration) or by tissue tests (semi-quantitative analysis) that frequently be practiced in the field.

5.2 Soil Testing

Soil test is considered quantitative analysis to evaluate the readily extractable nutrients in a soil. Soil tests can also make possible to assess the precise concentration of available nutrients, which plants may acquire from soil solution later. Usually, fertilizes recommendations are based on the interpretation of the results of the soil analysis. A fundamental concept of soil testing is that, specific points from a field are sampled to interpret the fertility status of the whole sampled area. The sampling methods may vary based on the inconsistency of the area being sampled and the nutrient under consideration.

A well representative soil test report literally relies on an accurately collected soil sample. For instance, if the field seems homogenous, one composite soil sample may be representative, whereas if the abnormal growth area exists, then one composite sample needed from the normal area and another from the problematic area. Furthermore, contemporary fertilization of a field can alter the spatial distribution of nutrient in a field, and to characterize the actual soil fertility level will become difficult. Additionally, variations in fertilizer application patterns, for example, placement of fertilizers in bands or broadcasting can influence soil samples. Therefore, the results of soil tests should be verified to crop responses, for instance, growth and yield with application of recommended fertilizers. Nevertheless, while calibrating soil test with crop response to applied fertilizers other crop yield variable limiting factors may also be considered, for instance, adverse environmental conditions, weeds, incidences of diseases or others insects pest.

It has been well documented that K in soils exists in various pools, namely water-soluble, exchangeable, non-exchangeable or fixed, and mineral K (see Chap. 2 for detail). Therefore, the interpretation of the K contents in soils requires some basic information about soil which is under consideration, for instance, clay contents, degree of soil weathering, and type and ratio of clay minerals. Nevertheless, if these all are identified, it remains ambiguous concerning what degree the interlayer K is mined and how much mica of the silt fraction participates in K availability for crops (Mengel et al. 1998). To date, much discussion has been carried out regarding a suitable method of extraction of K from the soil.

The most frequently used method for assessment of plant available K for recommendations of potash fertilizers especially in India and Pakistan is NH_4OAc-extractable K. In brief, K is extracted with 1 M NH_4 acetate at a pH 7. This method mainly represents total K in soil solution, exchangeable and small fraction of interlayer K but cannot distinguish their individual contribution. Collected results usually ranges between 40 and 400 mg kg^{-1} soil and sometime more than this value. The value fall below 120 mg K kg^{-1} soil are commonly categorized as K deficient, while value in the range of 120–250 mg K kg^{-1} soil are considered enough to support plant growth. The other method used is AB-DTPA, which also comprises exchangeable and water-soluble K; though, its value is comparatively lower than NH_4-acetate exchangeable K extraction method. Soil available K can also be extracted with 1 mM HCl, where the exchangeable K and small share of interlayer K are detached together. Nevertheless, the involvement of the interlayer K cannot be calculated. The same case is for soil extraction with 0.25 mM Ca lactate and HCl at a pH of 3.6 (Hoffmann 1991). The values of extracted K with this method are commonly to some extent lower from the actual value of exchangeable K fraction.

Besides, we can distinguish non-hydrated exchangeable and the interlayer K by electro-ultra-filtration (EUF) technique, through extraction of K from a soil suspension with provision specific electrical field (Schubert et al. 1989; Rao et al. 2000; Wakeel et al. 2013). The basic principle of EUF approach is presented in Fig. 5.2. In brief, extraction of the exchangeable K is set for 30 min (200 V, 20 °C); and for second fraction, the same soil suspension is extracted for further 5 min (400 V, 80 °C). The detached K during this 5-min period truly represent the soil interlayer K and that can be considered for precise recommendation of the potash fertilizers. This EUF extraction method is successfully adopted in Austria and Germany for field crops potash recommendation, especially the sugar beet. Furthermore, EUF technique is not only valuable for extraction of available K but it can be effectively used for other plant nutrients such as N, P, Mg, Ca, and micronutrients even in a one soil sample.

In brief, soils are highly heterogenous in nature, therefore, crop response to applied potash is generally inconsistent; even at enough soil K level plants might prone to K deficiency due to close association of K with soil mineralogy. Therefore,

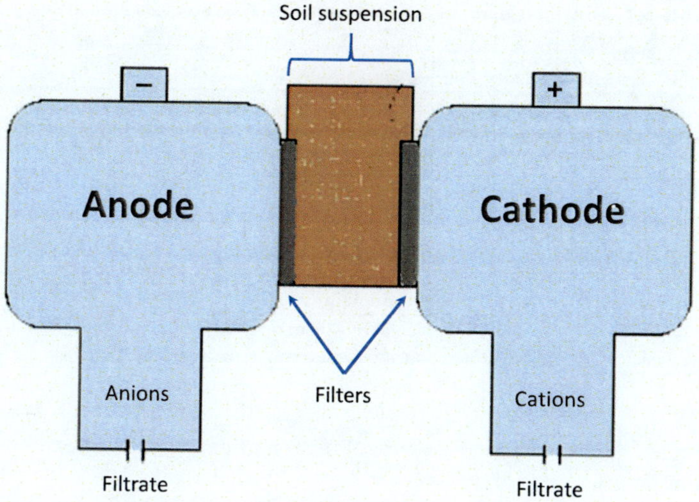

Fig. 5.2 Schematic principle of the electro ultra-filtration (EUF) apparatus. Migration of ions to electrodes result in dilution of ions in the solution and thus promotes the desorption of adsorbed ions (adapted from Mengel and Uhlenbecker 1993)

while recommending potash containing fertilizers, soil K status along with K-fixing capacity and mineralogical composition of soil must be taken into consideration.

5.3 Plant Tissue Analysis

Plant chemical analysis can be an additional valuable approach to distinguish variations in plant growth or composition that are synergistic or antagonistic with nutrient application. Furthermore, with appropriate practice of partitioning of individual constituents, such quantitative analysis may distinguish nutrients that are existing as soluble part of plant sap or have been assimilated into structures. In addition, plant analysis individually can make a satisfactory fertilizer recommendation for different crop species, for instance, fruit trees and vegetables. Furthermore, it is also distinguished tool to detect hidden hunger of distinct nutrient or incipient deficiencies. Hence, plant and soil analysis are decisive tools to authenticate soil nutrient availability and stress of the nutrients in plant.

Evaluation can be made between calculated data from plant samples with critical ranges that can be categorized as deficient, low, sufficient, or high concentration. The critical value of any nutrient in plant can be considered as the specific level of the nutrient below which yield of the crop tend to reduce. Besides, the actual demand for the specific nutrient may differ if other nutrients are not available above their own critical concentrations. Nutrient interactions, namely synergistic or antagonistic, can enhance or reduce the uptake of other nutrients via plant roots. Sometimes, plant growth may be depended on a nutrient up to degree that another nutrient become

Table 5.1 Distinct features of different tools of potassium evaluation in soil

Evaluation tool	Category	Remarks
Visual observation	Non-quantitative	K deficiency symptoms can be misdiagnosed with other nutrients or stresses, irreversible if not identified at earlier stages, cannot identify mild deficiency or "hidden hunger"
Soil analysis	Quantitative	Soil mineralogical composition, degree of weathering, and the extent the exhausted interlayer K are needed to know for precise evaluation
Plant analysis	Quantitative	Provide actual status of K in plant can be expensive, and time consuming
Tissue analysis	Semi-quantitative	Valuable under field condition and quick tool to diagnose K status or incipient nutrient deficiencies

limited, and additional supply of first nutrient cannot affect positively (law of minimum).

Therefore, the combination of soil as well as plant analysis is a finest approach to detect deficiency symptoms of the nutrient and subsequently fertilizer recommendation (Table 5.1). The selection of the plant part for analysis differs regarding nutrient going to be analyzed. However, K is mobile element which exist in soluble part of plant sap; therefore, the right choice of plant part is not as such considerable as compared with other nutrients such as nitrate (NO_3^-). All plant parts such as leaf petioles, blades, midribs, or other tissues can be selected for K analysis (Huang et al. 1992).

Potassium concentration in plant is determined usually by two methods, namely dry-ashing and wet digestion methods. Dry-ashing is a simple digestion method, non-hazardous, and valuable for low silica content crops, for instance, legumes. On the other hand, di-acid (HNO_3-$HClO_4$) or H_2SO_4 is used for wet digestion for full recovery of K in prepared plant samples, particularly containing high silica content, for example, wheat, rice, barley, sugarcane, etc. In brief, after digestion of the plant sample at standard temperature, the concentration of K can be easily detected by inductively coupled plasma atomic emission spectrometer (ICP-AES), atomic absorption spectrophotometer (AAS) or flame photometer. The K concentration in plant varies depending on plant species, growth stage, and other management practices (see Chap. 3 for detail description).

Plant tissue analysis is another technique that is basically semi-quantitative analysis tool and considerably valuable for growing plants. It is an approach for the quick determination of the nutritional status and is often directed in the field sites. In this method, nutritional status of the plant is usually assessed by direct determining the unassimilated soluble fraction of the nutrient. In short, in a plant analysis method, the sap of the tissues of relevant plant is extracted through methods, i.e. crushing or grinding together with filtering to get liquid for analysis. In addition, it can be valuable to diagnose an imminent nutrient deficiency, and it represents the actual availability of any specific nutrient (Imakumbili et al. 2020). In field crops, plant tissue testing along with soil analysis can assist precise fertilizer recommendation according to crop requirements. Analysis of different parts of plant is appreciated approach for crop management practices.

In conclusion, various environmental and edaphic factors influence the soil fertility and availability of the nutrients for plants. The edaphic factors that can affect the K availability can be soil mineralogical composition, the degree of weathering, soil pH, CEC, temperature, moisture contents, etc. In addition, K deficiency symptoms can be misdiagnosed with other biotic factors, namely disease, insect attack or herbicide damage, etc. Therefore, the application of potash fertilizer without considering all these factors may not be valuable to attain desirable yield and use of soil analysis and plant/plant tissue analyses should be considered for precise recommendations.

References

Barnes J, Whipker B, McCall I, Frantz J (2012) Nutrient disorders of 'evolution' mealy-cup sage. HortTechnology 22:502–508

Cakmak I, Kirkby EA (2008) Role of magnesium in carbon partitioning and alleviating photooxidative damage. Physiol Plant 133:692–704

Hoffmann G (1991) Die Untersuchung von Böden. Methodenbuch, vierte Auflage. VDLUFA-Verlag, Darmstadt, Germany

Huang WZ, Liang XY, Lun XJ (1992) Diagnosis of potassium deficiency in bananas using the method of different values. Commun Soil Sci Plant Anal 23:75–84

Imakumbili MLE, Semu E, Semoka JMR, Abass A, Mkamilo G (2020) Plant tissue analysis as a tool for predicting fertiliser needs for low cyanogenic glucoside levels in cassava roots: an assessment of its possible use. PLoS One 15:e0228641

Mengel K, Uhlenbecker K (1993) Determination of available interlayer potassium and its uptake ryegrass. Soil Sci Soc Am J 11:206–225

Mengel K, Rahmatullah, Dou H (1998) Release of potassium from the silt and sand fraction of loess-derived soils. Soil Sci 163:805–813

Rao CS, Subba RA, Swarup A, Bansal SK, Rajagopal V (2000) Monitoring the changes in soil potassium by extraction procedures and electroultrafiltration (EUF) in a Tropaquept under twenty years of rice-rice cropping. Nutr Cycl Agroecosyst 56:277–282

Schubert S, Paul R, Uhlenbecker K (1989) Charakterisierung des nachlieferbaren Kaliums aus der nichtaustauchbaren Fraktion von acht Böden mittels einer Austauschermethode und EUF. VDLUFA Schriftenreihe 30:329–334

Tanoi K, Kobayashi NI (2015) Leaf senescence by magnesium deficiency. Plan Theory 4:756–772

Verbruggen N, Hermans C (2013) Physiological and molecular responses to magnesium nutritional imbalance in plants. Plant Soil 368:87–99

Wakeel A, Gul M, Sanaullah M (2013) Potassium dynamics in three alluvial soils differing in clay contents. Emirates J Food Agric 25:39–44

6 Potash for Quality of Agricultural Commodities

Potash is equally important for all types of crops; however, the amount of K required may be different for each crop, and K acquisition from soil may also vary. Furthermore, quality conscious crops may require more sustainable potash supply. Therefore, quality aspects and K nutrition have been discussed for various crops separately in the following contents.

6.1 Cereals

Potassium has been acknowledged as *"quality element"*, which not only maintain desirable crop yield, but also ensure high-quality produce for human consumption. To a large extent, it participates in the production and translocation of photo-assimilates (sucrose) from the source to sink organs in other words from fully expanded mature leaves to grains. In addition, it has specific role in the activation of numerous enzymes that are essential for metabolic processes, particularly the synthesis of proteins. Meanwhile, K sustains the water content and accordingly the turgor pressure of cells which facilitate the leaf's vigor to proceeds photosynthesis efficiently. Such specific roles of K, for instance, water and nutrient uptake and transport within plant, along with the translocation of photosynthates produced by the photosynthesis process to storage organs improve the produce quality. When K is limited supply, photosynthesis, respiration, sugar translocation, and many enzymes activity impaired (Fig. 6.1).

The role of adequate K supply in physiological processes to improve the quality of crops is well recognized (Kumar et al. 2006; Pettigrew 2008). In brief, it mediates to accomplish specific physiological processes, i.e. increasing photosynthesis efficiency, increasing leaf surface area, and more efficient photosynthates and amino-N compounds translocation towards reproductive organs through the phloem (Cakmak 2005; Pettigrew 2008). Findings from various research studies regarding the potash nutrition in improving crop quality gives impression that much greater efforts are

A. Wakeel, M. Ishfaq, *Potash Use and Dynamics in Agriculture*,
https://doi.org/10.1007/978-981-16-6883-8_6

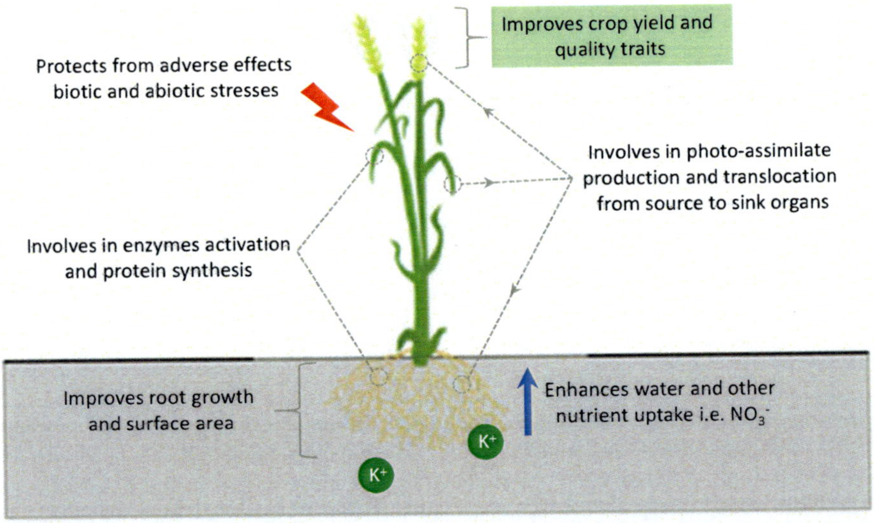

Fig. 6.1 Potassium is known as "quality element" as a result of its imperative effects on size, shape, color, taste, shelf life, nutritional value, and other quality traits. In addition to enzymes activation and protein synthesis, the translocation of photosynthates from the source to sink organs, i.e. roots, grain, fruit, or any other edible portion are supposed to be the key contributors responsible to improve crop yield and quality attributes mediated by adequate K supply

needed to assure that farming community more familiar to the consequences of enough K status in crop plants. Improving K nutrition can enhance both quantity and quality of agriculture commodities, largely under adverse conditions such as drought and salinity.

As K is a central element in many physicochemical processes within the plant and its deficiency is likely to reduce the efficiency of these processes, consequently, adversely affect crop performance and output quality. Reduction in the transport of photo-assimilates to the grain under K and/or magnesium (Mg) negatively effect on both yield and quality traits. With inadequate K supply, the breakdown of premature parenchyma had been noticed even when microscopic observation exposed no pathogenic organisms. Therefore, an adequate K availability is indispensable to the quality of the harvested crop. However, the amount of potash needed to attain desirable yields and quality differs regarding growth and environmental conditions.

Quality aspects of cereals comprise grain protein, starch, oil content, stalk quality and standability, crop maturation (moisture of grain), and seed quality as determined by sizing, coloration, and disease. In general, the quantity of K required to obtain maximum yield is also adequate to attain good quality. For instance, the favorable effect of K nutrition on the protein content has been testified in wheat and barley (Holland et al. 2019). Furthermore, combined application of K and salicylic acid under drought stress condition enhanced the nutritional value and water-soluble pentosan that mediates the bread-making quality (Safar-Noori et al. 2018).

The positive correlation in potash fertilization and wheat grain protein, harvestability, milling, and baking qualities have been revealed in different research studies. It facilitates N absorption, translocation of amino acid to developing grain, and accordingly improves gluten and prolamine synthesis, in addition to the synthesis of proteins, which improve baking quality of wheat. Wheat field trials conducted in India with K adequate supply revealed the better seed appearance than control, i.e. bolder and shinier. Additionally, K application increases the starch content in cereal crops such as rice, wheat, and corn. It is testified that rice yielded more with K fertilization, because of a longer period of grain filling and hence larger seeds that is more thousand-grain weight.

The beneficial aspects of K supply are well appreciated in corn, i.e. stalk strength, disease resistance, and less harvest loss as well as earlier silking, longer grain fill, uniform maturity, improve stock quality, reduced lodging, increase kernel yield, and more efficient use of N. The balance application of N and K, and other management practices can improve higher yields, protein, and starch content in maize (Jiang et al. 2014; Li and Shi 2016). Furthermore, damage from disease, insects, and environmental stresses are frequently reduced with adequate K availability. When maize was grown without manure application in China, a greater grain protein content was produced using a balanced NPK fertilizer rather than using a fertilizer containing only N and P (Yang et al. 2004).

As humans regularly consume a diet rich in cereal-based products, therefore, satisfactory K supply endorses the formation of grain which are larger in size with more protein and K contents. Nevertheless, with increasing consumption of processed food and K-depleted cereals, in addition to reduction in the K-rich vegetables and fruits consumption, K intake has been largely reduced which is now common in developed countries. A number of clinical and epidemiological research express that an improved K intake has advantageous influences on human health (see Chap. 7). Therefore, to improve the K intake through cereal grains, the efficient and higher potash nutrition is mandatory (Pettigrew 2008).

6.2 Sugar Crops

As it is explained earlier, K mediates the production and translocation of photosynthates from the source to sink organs therefore, its deficiency lead to the accumulation of sucrose in the source leaves. Furthermore, K also involves in photosynthesis rate, leaf development, and stress resistance, i.e. drought and diseases which all ultimately improve quality traits. The significance of K in the yield and quality of sugar crops, namely sugarcane (*Saccharum officinarum* L.) and sugar beet (*Beta vulgaris* L.) are closely studied. A balanced fertilization program with N and K, and the highest juice quality, sugar recovery (sweetness), and sugarcane yield is well documented in number of reports.

The quality traits of sugarcane with adequate K fertilization have been positively correlated. For example, K application considerably improved sucrose content in sugarcane (Khadar et al. 2004). A close relationship was experienced between the improvement in yield and percent increase in sucrose contents with the application of K under salt stress (Ashraf et al. 2009). In another study, significantly improvement in sucrose contents was detected with potash fertilization in sugarcane. Additionally, natural potash application from seaweed improved yield and quality of cane and cane juice (Karthikeyan and Shanmugam 2017). In the field trials during 2005 to 2008 piloted by Engro Chemical Pakistan Limited (ECPL) on sugarcane and potato in rice and central zones of Pakistan showed that with K application both these crops respond positively in the term of yield, whereas sugarcane quality measured as sucrose contents also increased (Ahmad and Mian 2009).

The fundamental roles of K in photosynthesis, sugar translocation, biotic and abiotic resistance, and lodging tolerance are crucial for high yield and quality of sugarcane. Appropriate nutrient management and different sources of potash fertilizer are imperative to improve sugarcane quality. For instance, a study of comparative application of KCl and K_2SO_4 disclosed that they have diverse effects on the quality of sugarcane and over application of KCl can reduce sucrose concentration in juice, specifying that Cl^- is the crucial factor for the quality development (Watanabe et al. 2016). Therefore, adverse effects of Cl^- on the quality of sugarcane can be reduced if it is not applied above sufficiency levels (Watanabe et al. 2017). Furthermore, fertilization program with combined N and K application have been revealed to guarantee both the maximum juice quality along with sugarcane yield.

Sugar beet is another important sugar crop due to its short duration (5–6 months) and high sucrose content (14–20%). It is well recognized that supply of optimum K promotes higher root yield with improved sugar content and accordingly higher industrial sugar yield (Mubarak et al. 2016). In contrast, under limited K supply reduced translocation of sucrose from the leaves to the sink, namely beet, due to impaired phloem loading reducing beet yield and quality (Fig. 6.1). Besides restricted sugar translocation, beet cultivated with inadequate K and high N nutrition has higher concentration of amino-N, which after harvesting reduces the extractability of sugar from the beet due to mollassegenic effect. The conversion of amino-N to protein reduces under limited potash supply (see Chap. 11) and an increase root, top, and sugar yields as well as TSS% were noticed when concentration of N and K was increased (Khaled et al. 2019).

The quality traits of sugar beets mediated by K supply can be assessed by measurements of juice purity, and refined sugar content. Under inadequate K nutrition, a decline can be observed in the purity coefficient of juice thickness. Meanwhile, K application can improve total soluble solids, soluble sugar content, purity percentage, and white sugar in beet (Mehrandish et al. 2012). With higher N supply, both amino acid content and alkalinity influence the sugar extractability. On the other hand, improved potash in relationship with N fertilization can improve the yield of refined sugar lead to a corresponding reduction in sugar content of molasses. Furthermore, in 2 years field experiment, it was revealed that foliar application of K and Si improved sugar content of tubers about 0.6–1.13% on the average of both years in sugar beet (Hrivna et al. 2017). However, a distinct relationship between soil type and beet sugar level mediated by K nutrition is poorly understood.

6.3 Vegetables

The quality of vegetables is a complex trait which largely composed of appearance or attractiveness, uniformity of size, nutritional value (i.e. proteins and vitamin C contents), attractive color, shelf life, injury free, markings, and indications of disease. Although quantity is the primary element driving the producer's price premium, nevertheless quality of product is preferable for consumers, therefore, often sets as a standard for the market price. A linear relationship is observed between the soil and/or foliar application of K with the level of vitamin C, flavor, and color in different vegetables such as cucurbits, cauliflower, and onion. The limited K supply in potatoes, tomatoes, and cabbage often direct to discoloration of internal tissue. Both tomato and potato respond well to K nutrition in the context of yield and percent of that yield fulfilling strict market standards. In addition, adequate K level also sustain quality during storage and transportation, hence extend shelf life.

Potassium application considerably improves quality traits of vegetables predominantly color value. Available N and K nutritions have substantial positive correlation with quality traits in chilies. For instance, pungency and the red color of chilies are influenced by N and K nutrition, respectively. However, with the application of different sources and level of potash, highest color value and oleoresin content were observed at 60 kg ha^{-1} in response to sulfate of potash application (Ananthi et al. 2004). Furthermore, significant improvement in ascorbic acid content has been observed with the supply of K (Majumdar et al. 2000). Soil types also influence the quality of red chilies, i.e. capsaicin and color value (Bidari et al. 2004). Furthermore, split application of K in chilies is revealed advantageous to produce attractive bright red colored fruits. For instance, potash fertilization in split doses verified maximum ascorbic acid content, color value, and oleoresin content in red chilies (Prabhavathi et al. 2008). It is revealed that K application improved vitamin C content (Wu-Zhong et al. 2001), fruit firmness, soluble sugars, TSS content, and ascorbic acid concentration in pepper (Botella et al. 2017).

In tomato, the development of red colored fruit is because of numerous internal pigments, particularly lycopene content that depends on the availability of K. The yield, quality, shelf life, total solids, acids, carotene, sugars, lycopene, and vitamin C in tomato fruit increases with enough K supply. In contrast, K deficiency lead to uneven ripening, yellow shoulder fruit, and irregularly shaped fruits with poor internal quality. Therefore, the reducing the appearance of tomato yellow shoulder requires an adequate supply of K in the soil (Hartz et al. 2005). Furthermore, potash fertilization also reduces the physiological disorders of tomato, for example, gold specks, blotchy ripening complex, puffiness, gray wall, and greenback, etc. The optimum K and calcium balance are positively correlated with tomato firmness, color characteristics, lycopene, and sugar concentrations (Hernández-Pérez et al. 2020). All these characteristics are important for economic value of the produce, especially considering high standard markets.

Potassium application considerably improved ascorbic acid content of brinjal (Thakre et al. 2005). Furthermore, the combined application of N and K improved sugar content, essential amino acids, and the total amount of amino acids in Chinese cabbage (Yang et al. 2019). The response of carrots with potash nutrition is measured as improvement in quality traits and retardation of wilting stress. Additionally, in sweet potatoes (*Ipomoea batatas* L.), potash application can improve tuber shape, starch content, decreased firmness of canned tubers, carotenoid content, and crude fiber.

Potassium is imperative in numerous physiological processes that mediates the potato (*Solanum tuberosum* L.) tuber quality, for instance, adequate rate of K supply improved tuber yield and quality traits including starch, protein and total carbohydrate contents (Elkhatib et al. 2019). At elevated K level, potato use less water per kg of tubers, survive drought periods and resist to external stresses, for example, heat impact, frost, and diseases, and consequently improve quality traits of tuber. Furthermore, adequate K supply lead to improve potato storage properties and reduces bruising damage that usually occurs during handling and transportation. Enough K level also promotes uniform light-colored chips by reducing chemical compounds that darken the chips during cooking. In contrast, inadequate K nutrition can reduce tuber dry matter by retarding photosynthesis activity, and henceforth less starch production. Additionally, internal blackening of potato tubers may be due to an additional of tyrosine produced under limited K availability.

The whole range of quality criteria can influence with availability of K during vegetables growing. Under optimum K nutrition, its concentration in potato lies nearly 4 g kg^{-1} dry matter for human consumption (Belitz et al. 2007). As potash application to potato can substantially influence the concentration of reducing sugars, which are central precursors for the acrylamide formation during frying. Furthermore, potato cultivated with low K and high N supply can accumulate most precursors for acrylamide formation during frying (Gerendás et al. 2007). The additional K supply can elevate malate and citrate concentrations, while reduces the concentrations of reducing sugars (Gerendás et al. 2007).

6.4 Fruits

The quality defining traits of fruits vary and sometimes need to improve fruit quality is more valuable than other aspects of yield production, particularly when high quality secures the satisfactory economic return. Fruit quality has many characteristics and the most specific aspects of quality depends on the fruit, for instance, it may be the thickness of a citrus peel and Vitamin C concentration or sugar concentrations are especially important for apples. In this context, more K supply is needed to ensure quality attributes than for maximum yield. Research endures to assess the consequences of K on number of quality traits and best marketable price premium. The general fruit quality traits include size, appearance, rind thickness, coarseness of peel, fruit color, soluble solids of fruit, acid/sugar ratio, texture, vitamin C content, and other nutritional value.

Potassium is essential for sustaining both the yield and quality of numerous fruits. The positive correlation of K supply and shelf life of fruit are observed, by slowing senescence process or reduction in numerous physiological diseases. As K improves

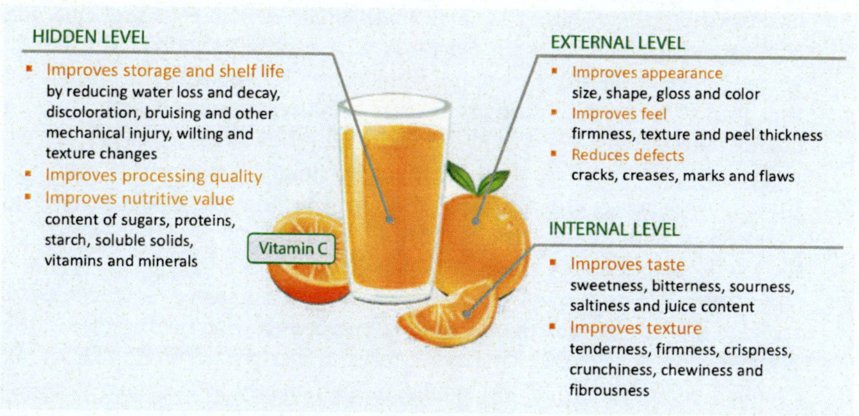

Fig. 6.2 Potassium has been termed as the *"quality element,"* which ensures the optimum quality of agricultural commodities at numerous levels. It lowers production costs and reduces the risk of rejection when the product is offered for sale (modified from IPI infographic)

strength and firmness the fruits skin, therefore they are less susceptible to impairment during transportation, resist to deterioration for a longer period and stay fresh (Kumar et al. 2006). Furthermore, potash fertilization decreases post-harvest moisture loss by maintaining tissue integrity and improving the weight of the harvested organs. As storage compounds accumulated in the harvested fruits accelerate few metabolic activities during storage. For example, organic acids and starch into CO_2 and water or respiration comprises the oxidative breakdown of sugars and K deficiency enhances the respiration rate. Furthermore, K also diminishes the incidence of some storage fungal diseases that can cause considerably losses, because minor damage in fruit must be discarded before marketing.

Fruit crops receiving enough K exhibits larger root development with much larger sugar content in fruits. Optimum K supply has been allied with improved fruit size, yields, ascorbic acid concentrations, and soluble solids as well as fruit shelf life, color, and shipping quality of several horticultural crops (Lester et al. 2007; Kanai et al. 2007). Meanwhile, inadequate K supply results in reduced translocation of photosynthates (sucrose) from the source into the fruits owing to impaired phloem loading (Fig. 6.1). In many fruiting crops, various research studies were conducted to compare soil *vs.* foliar application of potash, and foliar application of K consistently reported in more improved fruit quality traits. Furthermore, potash fertilization through both approaches revealed to improve the level of ascorbic acid in banana, papaya, and guava.

In citrus, acceptable potash fertilization can enhance yields, size, color, thickness of the rind (Lester et al. 2010), concentration of vitamin C, ascorbic and citric acid content of the juice, in addition to other juice qualities, such as soluble solids content and the acid/sugar ratio (Fig. 6.2.). The improved yield with K supply is due to decreased fruit fall from the tree and larger fruit size. Additionally, fruit rot is observed to reduce during storage while larger shelf life in the supermarket. In contrast, K deficiency, which outcomes in thin-skinned and small fruit size along with high rate of premature fruit fall. Under sever K deficiency, deterioration of fresh fruit lead to larger loss during transport and a reduced shelf life. The citrus fruit quality during storage is encouraged with K nutrition as the incidence of green mold and stem-end rot declines with improving K supply, consequently during transport fruit loss decrease and its shelf life improve. Furthermore, the physiological disorders of citrus fruits such as creasing and plugging are allied with high N and the low K supply during fruit development.

Grapevine (*Vitis vinifera*) is a valuable drought resistance plant and K intermediate vital role for its growth and yield in addition to wine quality (Mpelasoka et al. 2003; Villette et al. 2020). Grapevine acts as a strong sink for K, VvK3.1, an AKT2 channel, control berry K^+ accumulation, particularly at the time of ripening and improve its taste (Nieves-Cordones et al. 2019). The role of K is not much investigated in berries. In a recent study molecular study exposed that voltage-gated inwardly rectifying K channel (VvK1.2) was upregulated upon acidification of the berry (Cuéllar et al. 2013). This channel is particularly located in the berry, where it participates as prominent induction at the beginning of ripening. Furthermore, K contents are recorded larger in the skin of the berry, where anthocyanins are located, as compared to the fleshy pericarp. Consequently, K availability is valuable for the fermentation of red wine.

The yield and quality of banana are also largely attributed by K supply, and it improves fruit number, weight per bunch, visual appearance, soluble solids, sugar to acid ratio, and shelf life (Nikagolla et al. 2019). However, the positive influences of K nutrition on banana fruit quality continue in addition to the concentration of K required for optimum fruit yield. Meanwhile, limited K supply lead to smaller fruit growing on fragile and thin bunches, with a shorter shelf life in the market. Additionally, potash availability at initial stages of fruit development is correlated with increase in fruit number, weight, and fingers per bunch (Lorena et al. 2016). Furthermore, K promotes earlier fruiting and decreased the number of days to fruit maturity. In contrast, inadequate K lead to thin and brittle bunches with a reduced shelf life of banana. In grapes, K deficiency is also closely associated with reduced firmness and less juice.

Adequate K enhances color, TSS, appearance, and flesh firmness of apples by minimizing cell breakdown during storage (Nava et al. 2008; Chen et al. 2016). It is tested that internal apple fruit breakdown was responsible due to low levels of K. The combined application of N and K is supposed to valuable for both fruit firmness and color. It is revealed that N/K ratio of 1.25 is appropriate to produce desirable fruit quality of apple. Furthermore, the most imperative physiological post-harvest disorder, i.e. incidence of endogenous brown rot in pineapple can be reduced by supplying adequate amounts of K, which enhances the level of ascorbic acid, preventing the oxidation of phenolic compounds, and ultimately reduces the development of the disorder in fruits.

References

Ahmad Z, Mian SM (2009) Increasing crop productivity through judicious use of potash in Pakistan. Proceedings IPI-OUAT-IPNI International Symposium

Ananthi S, Veeraragavathatham D, Srinivasan K (2004) Influence of sources and levels of potassium on quality attributes of chilli (*Capsicum annuum* L.). South Indian Hort 52(1–6):152–157

Ashraf M, Rahmatullah R, Ahmad M, Afzal MA, Tahir SK, Maqsood MA (2009) Potassium and silicon improve yield and juice quality in sugarcane (*Saccharum officinarum* L.) under salt stress. J Agronomy Crop Sci 195:284–291

Belitz HD, Grosch W, Schieberle P (2007) Lehrbuch der Lebensmittelchemie. Springer, Heidelberg

Bidari, B.I., Martur, M.D. and Math, K.K. 2004. Influence of soil properties on yield and quality of chillies (*Capsicum annuum* L.) and partitioning of nutrients in fruit components. Paper presented in "Nation. Symp. on Input Use Efficiency in Agric. – Issues and Strategies", 25–27th November 2004, Kerala Agricultural University, Thrissur, Kerala

Botella MÁ, Arévalo L, Mestre TC, Rubio F, García-Sánchez F, Rivero RM, Martínez V (2017) Potassium fertilization enhances pepper fruit quality. J Plant Nutr 40(2):145–155

Cakmak I (2005) The role of potassium in alleviating detrimental effects of abiotic stresses in plants. J Plant Nutr Soil Sci 168:521–530

Chen C, Tong Y, Lu Y, Gao Y (2016) Effects of different potassium fertilizers on production, quality and storability of Fuji apple. J Plant Nutr Fert 22:216–224

Cuéllar T, Azeem F, Andrianteranagna M, Pascaud F, Verdeil JL, Sentenac H, Zimmermann S, Gaillard I (2013) Potassium transport in developing fleshy fruits: the grapevine inward K^+ channel VvK1. 2 is activated by CIPK–CBL complexes and induced in ripening berry flesh cells. Plant J 73(6):1006–1018

Elkhatib HA, Gabr SM, Mohamed RA, Manaa AF (2019) Impact of potassium fertilization rates and potassium solubilizing bacteria inoculation on the growth, yield and quality of potato (*Solanum tuberosum* L.). Alexandria Sci Exchange 40:127–142

Gerendás J, Heuser F, Sattelmacher B (2007) Influence of nitrogen and potassium supply on contents of acrylamide precursors in potato trubers and on acrylamide accumulation in French fries. J Plant Nutr 30:499–516

Hartz TK, Johnstone PR, Francis DM, Miyao EM (2005) Processing tomato yield and fruit quality improved with potassium fertigation. HortScience 40(6):1862–1867

Hernández-Pérez OI, Luis A, Valdez-Aguilar IAT, Andrew D, Cartmill and Donita L.C. (2020) Tomato fruit yield, quality, and nutrient status in response to potassium: calcium balance and electrical conductivity in the nutrient solution. J Soil Sci Plant Nutr 20:484–492

Holland JE, Cammarano D, Fitzgerald GJ, Perry EM, Poile G, Conyers MK (2019) Proximal fluorescence sensing of potassium responsive crops to develop improved predictions of biomass, yield and grain quality of wheat and barley. Precision Agric 20:379–397

Hrivna L, Kong JH, Machalkova L, Buresova I, Sapakova E, Kucerova J, Sottnikova V (2017) Effect of foliar nutrition of potassium and silicon on yield and quality of sugar beet in unusual windy conditions in 2014 and 2015. Listy Cukrovarnicke a Reparske 133:182–187

Jiang B, Pang Y, Yu Y, Wang C (2014) Effect of nitrogen to potassium ratio on dry matter accumulation, yield and quality of maize in cold area. J Maize Sci 22:137–142

Kanai S, Ohkura K, Adu-Gyamfi JJ, Mohapatra PK, Nguyen NT, Saneoka H, Fujita K (2007) J Exp Bot 58:2917–2928

Karthikeyan K, Shanmugam M (2017) The effect of potassium-rich biostimulant from seaweed Kappaphycus alvarezii on yield and quality of cane and cane juice of sugarcane var. Co 86032 under plantation and ratoon crops. J Appl Phycol 29:3245–3252

Khadar MS, Negm AY, Khalil FA, Antoun LW (2004). Effect of potassium chloride in comparison with potassium sulphate on sugar cane production and some soil chemical properties under Egyptian conditions. Regional workshop on Potassium and fertigation development in west Asia and North Africa, p 24–28.

Khaled AAA, Adel RY, El-Hag DA, Osman MF (2019) Effect of nitrogen and potassium fertilization on productivity and quality of sugar beet plants at north delta region. Fresenius Environ Bull 28:7833–7839

Kumar N, Kavino M, Kumar AR (2006) Balanced fertilization for sustainable yield and quality in tropical fruit crops. In: Benbi DK et al (eds) Balanced fertilization for sustaining crop productivity. Internat Potash Institute, Horgen, pp 387–405

Lester GE, Jifon JL, Stewart WM (2007) Foliar potassium improves cantaloupe marketable and nutritional quality. Better Crops 91(1):24–25

Lester GE, Jifon JL, Makus DJ (2010) Impact of potassium nutrition on food quality of fruits and vegetables: A condensed and concise review of the literature. Better Crops 94:18–21

Li B, Shi W (2016) Effects of different ratios of nitrogen, phosphorous and potassium on dry matter accumulation, yield and quality of maize. Jiangsu Agricult Sci 44:85–89

Lorena DR, Silva CLD, Peixoto JR, Oliveira GPD, Rios GFA (2016) Influence of irrigation and potassium fertilization on the production and quality of banana fruit in Brasilia. Eng Agric 24: 463–472

Majumdar SP, Meena RL, Baghel GDS (2000) Effect of levels of compaction and potassium on yield and quality of tomato and chilli crops grown on highly permeable soils. J Indian Soc Soil Sci 48(2):215–220

Mehrandish M, Moeini MJ, Armin M (2012) Sugar beet (*Beta vulgaris* L.) response to potassium application under full and deficit irrigation. Eur J Exp Biol 2:2113–2119

Mpelasoka BS, Schachtman DP, Treeby MT, Thomas MR (2003) A review of potassium nutrition in grapevines with special emphasis on berry accumulation. Aust J Grape Wine Res 9(3):154–168

Mubarak MU, Zahir M, Ahmad S, Wakeel A (2016) Sugar beet yield and industrial sugar contents improved by potassium fertilization under scarce and adequate moisture conditions. J Integr Agric 15(11):2620–2626

Nava G, Dechen AR, Nachtiga GR (2008) Nitrogen and potassium fertilization affect apple fruit quality in southern Brazil. Commun Soil Sci Plant Anal 39(1–2):96–107

Nieves-Cordones M, Andrianteranagna M, Cuéllar T, Chérel I, Gibrat R, Boeglin M (2019) Characterization of the grapevine Shaker K^+ channel VvK3.1 supports its function in massive potassium fluxes necessary for berry potassium loading and pulvinus-actuated leaf movements. New Phytol 222(1):286–300

Nikagolla N, Udugala-Ganehenege MY, Daundasekera WAM (2019) Postharvest application of potassium silicate improves keeping quality of banana. J Horticult Sci Biotechnol 94(6):735–743

Pettigrew WT (2008) Potassium influences on yield and quality production for maize, wheat, soybean and cotton. Physiol Plant 133:670–681

Prabhavathi K, Bidari BI, Shashidhara GB, Mathad JC (2008) Influence of sources and levels of potassium on quality attributes and nutrient composition of red chillies. Karnataka J Agricult Sci 21(3):379–381

Safar-Noori M, Assaha DVM, Saneoka H (2018) Effect of salicylic acid and potassium application on yield and grain nutritional quality of wheat under drought stress condition. Cereal Res Commun 46:558–568

Thakre CM, Badole WP, Tiwari TK, Sarode PB (2005) Effect of different levels of sulphur, phosphorus and potassium on yield and quality of brinjal. J Maharashtra Agric Univ 30(3):352–353

Villette J, Cuellar T, Verdeil JL, Delrot S, Gaillard I (2020) Grapevine potassium nutrition and fruit quality in the context of climate change. Front Plant Sci 11:123

Watanabe K, Fukuzawa Y, Kawasaki S, Ueno M, Kawamitsu Y (2016) Effects of potassium chloride and potassium sulfate on sucrose concentration in sugarcane juice under pot conditions. Sugar Tech 18(3):258–265

Watanabe K, Tominaga J, Yabuta S, Takaragawa H, Suwa R, Ueno M, Kawamitsu Y (2017) Effects of different kinds of potassium and chloride salts on sugarcane quality and photosynthesis. Sugar Tech 19(4):378–385

Wu-Zhong N, Liang JS, Härdter R (2001) Yield and quality responses of selected solanaceous vegetable crops to potassium fertilization. Pedosphere 11:251–255

Yang S, Fengmin L, Malhi SS, Wang P, Suo D, Wang J (2004) Long-term fertilization efforts on crop yield and nitrate nitrogen accumulation in soil in northwestern China. Agron J 96:1039–1049

Yang S, Chen Y, Yang D, Jin K, Tang J, Huang Y, Wang Z (2019) Effect of different nitrogen and potassium fertilizer combination on yield and quality of Chinese cabbage in two purple soils. Soil Fertilizer Sci China:115–121

7 Potassium Biofortification and Human Nutrition

Micronutrient malnutrition in humans is expanding worldwide and biofortification of trace elements to combat micronutrient malnutrition has been considered largely in different parts of world, especially the developing countries. Recent studies in the USA have also revealed K deficiency in the specific populations who are fed on agricultural commodities grown on K deficient soils. Therefore, K deficient soils in developing countries such as in Pakistan where K is not fertilized to the soils have greater chance of K deficiency in human population with great potential for biofortification of potassium.

7.1 Significance of Potassium for Human Health

Minerals are inorganic substances which mediate certain physical and chemical processes in all tissues and fluids, which are crucial to sustain healthy life. For humans, potassium (K) is the third most abundant micromineral after calcium (Ca) and phosphorus (P) (Lindshield 2018), and plants are the major source of K for human daily consumption. Within the human body, K is a fundamental cation in intracellular fluid (Calabrese et al. 2019), which contributes in the regulation of osmotic pressure, acid-base balance, conduction of nerve impulses, cell membrane function, muscle contraction, bones health, blood pressure, heart functions, maintain healthy body weight, and more (Weaver 2013; Fig. 7.1). While hypokalemia, or K deficiency, can lead to muscle weakness, fatigue, constipation, mental apathy, etc. The significance of K to human health has been well categorized, and innovative research findings persist to highlight its beneficial influences and possible consequence in human health.

A. Wakeel, M. Ishfaq, *Potash Use and Dynamics in Agriculture*,
https://doi.org/10.1007/978-981-16-6883-8_7

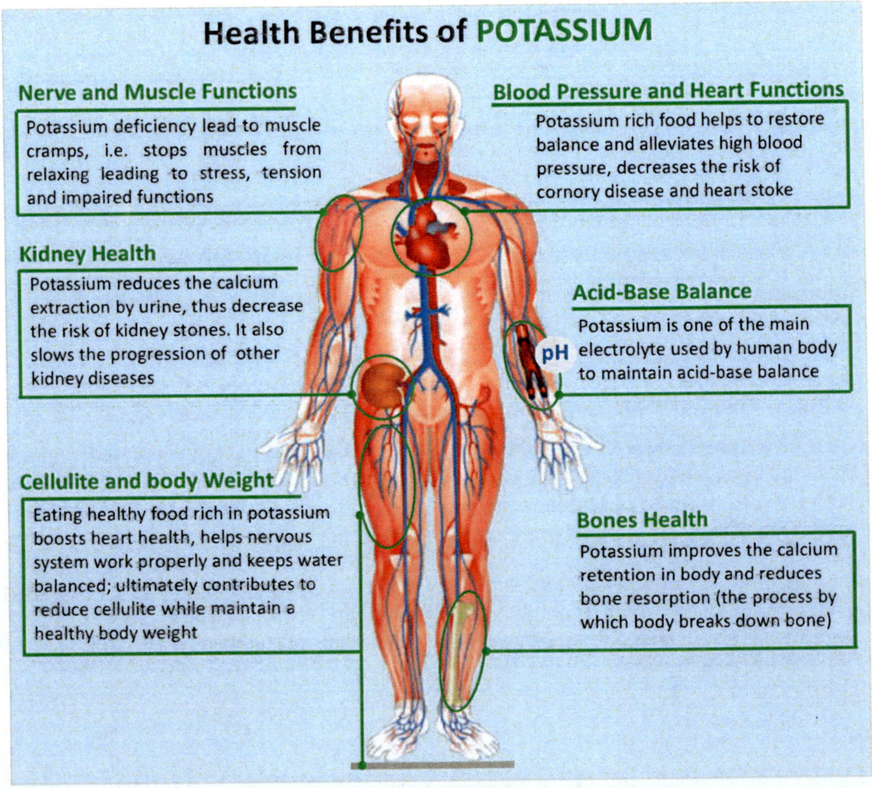

Fig. 7.1 Potassium is not only essential nutrient for plants, but its adequate intake is also compulsory to function human body appropriately. It is vital to maintain fluid balance, muscle contraction, conduction of nerve impulses, reduce the risk of stroke, and more. Lower intake of K can lead to high blood pressure, muscles cramps or weakness, and even can alter the proper heart rhythm

7.1.1 Blood Pressure

One of the leading beneficial aspects of adequate K nutrition is related to the regulation of blood pressure. Hypertension or blood pressure has been projected as a main cause for cardiovascular diseases, mainly heart stroke and coronary disease, which has been declared one of the foremost reasons of mortality worldwide. Improving dietary K intake is associated to alleviation of high blood pressure. A number of research findings have confirmed that improved K intake decreases systolic (maximum) and diastolic (minimum) blood pressure in both hypertensive (high blood pressure) and non-hypertensive patients. Nevertheless, the decreases in blood pressure is probable to higher in the patient who are hypertensive, which highlights the promising advantage of improving K intake in these patients. Data also proposes that the higher K level may regulate blood pressure, and consumption of K by foods containing enough K contents may apparently beneficial for most children (He and MacGregor 2008). An elevated K level also diminishes the undesirable consequences of sodium (Na) on blood pressure (Filippini et al. 2017). Meanwhile, food containing low K content, particularly having the elevated level of Na, has been concerned in the initiation of high blood pressure, which can consequently result in cardiovascular disease. Therefore, Na-K ratio of about 1:1 is considered appropriate to consume, nevertheless, as a matter of fact the ratio of Na to K literally intake by numerous people is as a minimum 2:1.

7.1.2 Cardiovascular Disease

An adequate K intake has also been associated in inhibiting other cardiovascular risks. So far, numerous research findings have revealed that improved K intake may reduce the probabilities of coronary artery disease and stroke. Many reports combined from different research findings also confirm these conclusions; it is revealed that every 1640 mg of K consumption per day can reduce the chances of stroke by 21%. On the other hand, wider Na–K ratio is correlated to higher consequence of cardiovascular diseases (Jayedi et al. 2019), ischemic heart disease, and higher mortality, which are considered as a result of the restricted blood supply to the heart. As mentioned in the previous sub-section, adequate K level control blood pressure, and may optimize the function of the cardiovascular system.

7.1.3 Bones Health

Improving K intake by vegetables and fruits is allied with bones health and development in children, adults, and the elderly. Research investigations on the influence of K nutrition on bone health have confirmed an affirmative correlation between higher K intake, and bone mass and bone mineral density. However, the precise process by which K improves the skeleton is still little elucidated and deliberated (Lambert et al. 2011), possibly improving K intake categorically decreases the urinary calcium (Ca) defecation and enhances Ca accumulation, consequently building a favorable calcium balance in body. Followed by, beneficial influence on bone mass and related risk of osteoporosis, which is a substantial human health issue throughout the world.

7.1.4 Kidney Health

The risk of kidney stones formation increases with aberrantly high urinary Ca (hypercalciuria). Therefore, by decreasing urinary Ca excretion through high K intake might diminish the risk of the development of kidney stones. Many research studies revealed that high K intake (nearly 4000 mg per day in men and 4700 mg per day in women) is related with decreased chances of kidney stones. Additionally, few findings also express that dietary K reduces the progression of kidney diseases (Kotnis et al. 2017; Morris et al. 2019). In hypertensive rats, adequate K intake was exposed to reduce adverse effects to the kidney, apart from its influence in the regulation of blood pressure. The K supplementation was also revealed to reduce renal inflammation in a rodent model of chronic kidney disease. Nevertheless, it remains open to further elucidate its impact in humans.

7.2 Potassium in Human Diet

The normal quantity of K in the human body is nearly 140 g, and our body demands a daily intake of K which is equivalent to around 2 kg K_2O per person per year. In 2004, the Food and Nutrition Board of the Institute of Medicine (2005) introduced that for male and female adequate level of K is 4700 mg per day (Table 7.1). However, in most regions of the world, dietary intake of K is reported lower than its recommended level. For instance, third National Health and Nutrition Examination Survey (NHANES III) data show that in the USA, the potash intake is adequate only in 10% of men and <1% of women. It is fairly low especially in the rural populations of developing countries where daily consumption of staple food with low K cereal products are dominant. Male has comparatively higher requirement of adequate intake of K than female and its recommended daily intake also increases with age.

The unrefined plant foods usually have higher concentration of K, especially fruits, grains, vegetables, and nuts which are the main source of K intake for humans. In addition, meat and milk are also rich sources of K nutrition. However, food processing diminishes the satisfactory K level in numerous food products (Karppanen et al. 2005). Thus, food commodities, which are usually processed before consumption, are often deficient in the adequate K level which is 4700 mg per day. In addition, during processing, Na is usually added, and K is removed which expand the Na–K ratio. Therefore, adequate K nutrition for human is imperative for health benefits especially under current expanding K deficient circumstances. Consumption of unprocessed, K-rich vegetables and fruits, dried fruits, avocado, meats,

Table 7.1 Adequate dietary intake of potassium recommendation (mg day^{-1})

Age group	Adequate intake level for K
0–6 months	400
7–12 months	700
1–3 years	3000
4–8 years	3800
9–13 years	4500
14–18 years	4700
>18 years	4700
Pregnancy (14–50 years)	4700
Lactation (14–50 years)	5100

Source: Food and Nutrition Board of the Institute of Medicine

fish, legumes, and the increasing K level in agriculture produce are the preferred and safest approach to improve K intake.

7.3 Importance of K Biofortification of Staple Crops

In earlier time, human's intake a diet ample in K, nevertheless, with the passage of time the consumption of processed foods and reduced consumption of vegetables and fruits increased, which ultimately reduced the adequate K level in our diet (He and MacGregor 2008). Furthermore, considering the tendency for increased consumption of K-rich vegetables and fruits with the increasing world's population, the quantity of soil potash is reducing in various agricultural farms. In addition, during the cooking process of various vegetables, there are chances to loss numerous mineral nutrients (Kimura and Itokawa 1990), in contrast, if fruits and vegetables are consumed as it is, for instance, consumption of green salad can minimize this issue. Therefore, it is obligatory to improve K contents in our diet, for instance, through biofortification program to assure balance diet for healthy life.

Biofortification of crops with application of essential mineral nutrients is considered as a direct approach which not only improves their desirable level in our food but also crop yield (Graham et al. 2007). Biofortification comprises of improving agronomic and nutritional value of the crop without transgenic approaches (Nestel et al. 2006). The saying "*Agriculture is the Guardian of Health*" *is* a fact, as the basis of human and animal health-relies on the supply of adequate amount of nutrients via agricultural commodities and soil is the starting point (Ishfaq et al. 2021). As few studies evidently confirmed a close correlation between K intake through diet and bone mineral density (Zhu et al. 2009). However, biofortification of staple crops with K application to improve K contents in our diet is very little documented. Much remains open to elucidate the possible potential of K biofortification for community health.

The application of potash containing fertilizers through soil and/or foliar spray can improve K contents in our diet. For instance, application of potassium sulfate in the range 0.014–4 g L^{-1} plus recommended level of other nutrients considerably improved K contents over time in cucumbers. The intake of K through human diet can be attained by improving K content in food crops and/or K supplementation to processed foods (Demingé et al. 2004; He and MacGregor 2008). Largely mineral nutrients are taken from different vegetables are not adequate, therefore most recent years biofortification is being considered a powerful tool for enriching vegetable products (Welch 2002). In brief, selecting K efficient genotypes along optimizing soil fertilization may be an advantageous approach for sustainable agriculture production (Rengel and Damon 2008) to improve human health.

References

Calabrese A, Gibby C, Meinke B, Revilla MKF, Titchenal A (2019) Human nutrition. Available at: http://pressbooks.library.ryerson.ca/vitalsign/

Demingé C, Sabboh H, Rémésy C, Meneton P (2004) Protective effects of high dietary potassium: nutritional and metabolic aspects. J Nutr 134:2903–2906

Filippini T, Violi F, D'Amico R, Vinceti M (2017) The effect of potassium supplementation on blood pressure in hypertensive subjects: a systematic review and meta-analysis. Int J Cardiol 230:127–135

Food and Nutrition Board, Institute of Medicine (2005) Potassium. Dietary reference intakes for water, potassium, sodium, chloride, and sulfate. National Academies Press, Washington, DC, pp 186–268

Graham RD, Welch RM, Saunders DA, Ortiz-Monasterio I, Bouis HE, Bonierbale M, De Haan S, Burgos G, Thiele G, Liria R, Meisner CA (2007) Nutritious subsistence food systems. Adv Agronomy 92:1–74

He FJ, MacGregor GA (2008) Beneficial effects of potassium on human health. Physiol Plantarum 133:725–735

Ishfaq M, Wakeel A, Shahzad MN, Kiran A, Li X (2021) Severity of zinc and iron malnutrition linked to low intake through a staple crop: a case study in east-Central Pakistan. Environ Geochem Health 43(10):4219–4233. https://doi.org/10.1007/s10653-021-00912-3

Jayedi A, Ghomashi F, Zargar MS, Shab-Bidar S (2019) Dietary sodium, sodium-to-potassium ratio, and risk of stroke: a systematic review and nonlinear dose-response meta-analysis. Clin Nutr 38:1092–1100

Karppanen H, Karppanen P, Mervaala E (2005) Why and how to echanges in food items and diets? J Human Hypertension 19:10–S19

Kimura M, Itokawa Y (1990) Cooking losses of minerals in foods and its nutritional significance. J Nutr Sci Vitaminol 36(4-SupplementI):25–S33

Kotnis A, Satyanarayana T, Majumdar K (2017) Vulnerability of human health due to inadequate dietary potassium is associated with potassium nutrition of crops. Indian J Fert 13(11):16–23

Lambert H, Boyd V, Darling A, Torgerson D, Burckhardt P, Frassetto L, Lanham-New S (2011) Dietary potassium and bone health: a systematic review and meta-analysis. Bone 48:S187–S203

Lindshield BL (2018) Kansas State University Human Nutrition (FNDH 400) Flexbook. NPP eBooks. 19. Available at: http://newprairiepress.org/ebooks/19

Morris A, Krishnan N, Kimani PK, Lycett D (2019) Effect of dietary potassium restriction on serum potassium, disease progression, and mortality in chronic kidney disease: a systematic review and meta-analysis. J Renal Nutr 30(4):276–285

Nestel P, Bouis HE, Meenakshi JV, Pfeiffer W (2006) Biofortification of staple food crops. J Nutr 136(4):1064–1067

Rengel Z, Damon PM (2008) Crops and genotypes differ in efficiency of potassium uptake and use. Physiol Plantarum 133:624–636

Weaver CM (2013) Potassium and health. Adv Nutr 4(3):368S–377S

Welch RM (2002) Breeding strategies for biofortified staple plant foods to reduce micronutrient malnutrition globally. J Nutr 132(3):495S–499S

Zhu K, Devine A, Prince RL (2009) The effects of high potassium consumption on bone mineral density in a prospective cohort study of elderly postmenopausal women. Osteoporosis Int 20: 335–340

8 Potash Research in Pakistan

Pakistan is an agricultural country and a major portion of its economy is dependent on agriculture; however, due to the lack of systematic research, the agriculture production is declining. Pakistani researchers in collaboration with overseas partners are working on potash since decades, however due to lack of systematic approach the work done could not influence the use of K fertilizer in Pakistan although researchers are in favor to enhance the use of potassic fertilizer in Pakistan agriculture. In this chapter research work done on K nutrition in Pakistani agriculture is compiled to give an outlook for researchers.

8.1 Crop Responses to Applied Potash

Potassium plays a key role in certain physiological and biochemical processes which ultimately influence crop growth, yield, and quality traits. However, crop responses to K supply are certainly variable because of different soil textures, which probably due to diverse clay contents and mineralogical composition. As a general rule of thumb, crop responses to applied potash are higher in crops particularly grown on soils which are low in K availability. Before the introduction of high yielding cultivars, the soils of Pakistan were not much exhausted as compared to recent times where several high yielding cultivars of various crops have been cultivated intensively. It led to deplete the soil K content due to inadequate replenishment or negligible use of potassic fertilizers and manures. Furthermore, mica based dominant soils have been transformed to illite and vermiculite rich clay minerals as a result of intensive cropping, continuous weathering, and accordingly soil mineral-K release (see Chap. 2). Therefore, K deficiency is evolving a limiting factor in agriculture production and adequate K supply is much needed for agricultural sustainability.

Recent research studies conducted in the range of agroclimatic conditions of Pakistan on various crops species support that largely soils are severe or marginal K deficient, and crop respond to applied potash in the context of growth and yield

A. Wakeel, M. Ishfaq, *Potash Use and Dynamics in Agriculture*,
https://doi.org/10.1007/978-981-16-6883-8_8

Table 8.1 Crop response to applied potash grown on soils of different regions of Pakistan using different fertilizer sources as well as fertilizer application rates

Crop	Soil K status	K source	Rate of K application	Application method	Increase over control	Reference
Cotton	153 mg K kg^{-1} soil	K_2SO_4	100 kg K_2O ha^{-1} soil, 2% foliar	Fertigation and foliar	8% lint yield, 15% seed-cotton yield with fertigation, and 15% lint yield, 22% seed-cotton yield with foliar than basal application	Hussain et al. (2020)
Cotton	163 mg K kg^{-1} soil	KCl	100 kg K_2O ha^{-1}	Soil	60–149% net income	Shahzad et al. (2019)
			200 kg K_2O ha^{-1}		57–157% net income	
Cotton	163 mg K kg^{-1} soil	K_2SO_4 and KNO_3	100 kg K_2O ha^{-1} + 2% 4 foliar spray	Soil plus foliar	59% intact fruit, 1-fold net income	Tariq et al. (2018)
			100 kg K_2O ha^{-1} + 2% 4 foliar spray		66% intact fruit, 92% net income	
Maize	92 mg K kg^{-1} soil	K_2SO_4	60 kg ha^{-1}	Soil	10% shoot growth	Khan et al. (2015)
			90 kg ha^{-1}		19% shoot growth	
			120 kg ha^{-1}		20% shoot growth	
Sugar beet	0.3 mmol L^{-1} soil solution	K_2SO_4	75 kg ha^{-1}	Soil	15% yield	Hussain et al. (2014)
			150 kg ha^{-1}		50% yield	
Sugarcane	–	K_2SO_4	42 kg ha^{-1}	Soil	73% yield	Soomro et al. (2014)
			84 kg ha^{-1}		101% yield	
			126 kg ha^{-1}		127% yield	
			168 kg ha^{-1}		190% yield	
			210 kg ha^{-1}		187% yield	
Tomato	97 mg K kg^{-1} soil	KNO_3	3.3 mmol kg^{-1} soil, 4.5 mM foliar	Soil and foliar	18% shoot growth when applied to soil, 12% shoot growth when applied as foliar spray	Amjad et al. (2014)
			6.6 mmol kg^{-1} soil, 9 mM foliar		27% shoot growth when applied to soil, 23% shoot growth when applied as foliar spray	
Wheat	0.3 mmol L^{-1} soil solution	K_2SO_4	50 kg K ha^{-1}	Soil	14% grain yield	Hussain et al. (2013)
			100 kg K ha^{-1}		30% grain yield	

Wheat	115 mg K kg^{-1} soil	K_2SO_4	80 g K_2O ha^{-1}	Soil	55% yield	Ashraf et al. (2011a, 2011b)
			120 kg K_2O ha^{-1}		160% yield	
Tomato	–	KNO_3	500 ppm	Foliar	39% shoot growth	Azeem and Ahmad (2011)
Cotton	–	K_2SO_4	30 mg pot^{-1}	Soil	8%, 34%, and 46% shoot K concentration under 70, 140, and 210 Mol m^{-3} NaCl stress, respectively	Akhtar et al. (2010)
Wheat	–	K_2SO_4	30 mg kg^{-1}	Soil	49% K concentration in leaf under salinity	Saqib et al. (2008)
Wheat	80 mg K kg^{-1} soil	K_2SO_4	60 kg ha^{-1}	Soil	13% yield	Khan et al. (2007)
Wheat	–	K_2SO_4	60 kg ha^{-1}	Soil	39% shoot K concentration at booting and milking stages	Majid et al. (2007)
Cotton	–	KCl	62.5 kg ha^{-1}	Soil	30% shoot growth	Pervaz et al. (2006)
			125.0 kg ha^{-1}		64% shoot growth	
			250.0 kg ha^{-1}		83% shoot growth	
		K_2SO_4	62.5 kg ha^{-1}		32% shoot growth	
			125.0 kg ha^{-1}		69% shoot growth	
			250.0 kg ha^{-1}		86% shoot growth	
Wheat	86 mg K kg^{-1} soil	KCl	75 kg ha^{-1}	Soil	6% shoot growth	Saifullah et al. (2002)
			150 kg ha^{-1}		16% shoot growth	
			225 kg ha^{-1}		29% shoot growth	
			300 kg ha^{-1}		27% shoot growth	
Wheat	–	K_2SO_4	60 kg ha^{-1}	Soil	12% shoot growth	Akhtar et al. (2000)

(Table 8.1). Certainly, there is still limited published data available regarding spatial crop responses to applied potash, especially in the arable farming of Balochistan and Sindh provinces of Pakistan. Pakistani farmers with limited financial resources usually do not apply fertilizer according to their soil fertility status or capacity of soil testing laboratories is limited to entertain all smallholder farmers. However, evidences regarding crop responses to applied K in Pakistan are expanding. Therefore, more precise research data is needed to emphasize the extent of K deficiency and crop response to applied potash fertilizers in distinct agroecosystem zones of Pakistan. Additionally, there are several limitations to adopt the balanced use of fertilizers in Pakistan that are comprehensively highlighted in Chap. 10.

8.2 Sources and Methods of Potash Application

Like nitrogen and phosphorus, K is also a major plasnt nutrient and is required in large amount to encompass certain physiochemical process in plants. There are numerous sources of potash fertilizers commercially available in the market, including sulfate of potash (SOP), muriate of potash (MOP/KCl), potassium nitrate (KNO_3), and polyhalite. Many of these include other plant nutrients, for example, magnesium and sulfur. The chemical composition and concentration of K in different available sources of potash are shown in Table 8.2. The major sources of potash adopted by local farmers are SOP and MOP, for instance, in cotton growing areas of Pakistan nearly 61% of farming community is applying SOP as a source of potash (Wakeel and Ishfaq 2016). Most recently more farmers are using MOP due to its high solubility, low price, and availability in the local market.

Potash fertilizers are predominantly water-soluble salts and available in ionic form in the soil solution or soil exchange sites. Potassium sulfate is preferred mainly for salt-sensitive crops in which chloride acquisition is problematic. It is sulfate-containing and non-hygroscopic in nature, nevertheless MOP has additional advantage of being cheap comparative to SOP. Furthermore, many field crops are not sensitive to chloride and can be fertilized with MOP using good quality irrigation water along with (see Chap. 11). As these both sources of potash are water soluble and do not affect soil pH. However, which source of potash is most appropriate

Table 8.2 Chemical composition and potassium concentration of commercially available potash containing fertilizers

Fertilizer	Chemical composition	K_2O (%)	Remarks
Muriate of potash	KCl	60	Contains chloride (48%)
Sulfate of potash	K_2SO_4	50	Contains Sulfur (18%)
Potassium nitrate	KNO_3	46	Contain nitrogen (13%)
Sulfate of potash magnesia	$K_2SO_4{\cdot}MgSO_4$	22	Contains S (21%) and mg (11%)
Polyhalite	$K_2Ca_2Mg(SO_4){\cdot}2H_2O$	14	4 nutrients in one single fertilizer (17% CaO, 6% MgO, and 48% SO_3)

under specific soil conditions, is still under debate. Differences in spatial-temporal variation along with economic resources and potential crop sensitivities are obligatory to recommend the right source of potash. Additionally, a new K source has been introduced in Potassium uptake by maize Pakistan named as polyhalite. Polyhalite is a cocktail of many nutrients containing K, Ca, Mg, and Sulfate. Slow release of nutrients and availability of Ca and Mg along with sulfate can be a potentially successful fertilizer in the soils of Pakistan. It is a natural fertilizer and approved for organic farming.

Time of optimum K uptake varies with crop species; however, plants usually accumulate maximum K at their earlier growth stages until flowering, after which it declines toward maturity. However, leaching especially in coarse-textured soils can be a reason for low K use efficiency (KUE) (Annadurai et al. 2000). Therefore, one approach to enhance KUE is its split applications during the crop growing season, which can minimize the leaching losses of K. For instance, growth and yield attributes of rice observed higher receiving split (half at transplanting and half at maximum tillering) application of K over whole basal application. Furthermore, soils having cation exchange capacity (CEC) medium to high and with 2:1 type clay minerology, (i.e. Alfisols, Inceptisols, Vertisols, and Mollisols) K can be supplied in all seasons with lesser susceptible to leaching. Meanwhile, Ultisols, Spodosols, Oxisols, and Histosols soils are prone to K leaching, especially during monsoon rainfall or in winter season. Therefore, K fertilizer can be applied in split doses during vegetative growth in order to reduce K leaching especially soil having low CEC. Furthermore, split application is also valuable approach to improve K uptake before it is subjected to fixation by soil.

Potash application in a band with a seeder or deep-banded usually enhanced its uptake via plant roots. Furthermore, banding potash fertilizer reported yield advantage over its surface broadcast application methods exclusively in soil with lower K status and crop residues. Furthermore, in agroecosystem having limited availability of irrigation water or rainfall, crop yields can be improved with deeper placement of K fertilizer. As localized placement may improve its availability to root vicinity, therefore, plant roots can efficiently take it up over surface application under limited water supply. Additionally, K is a monovalent cation (K^+) and readily adsorbed with soil colloids, therefore, it is not studied to be mobile in soil as nitrate (NO_3^-). Thus, plants readily take the K^+ up via roots which are in close proximity to the roots. Furthermore, crop distinct root morphology, especially the density of lateral roots has great significance in uptake of applied K.

Foliar application is another environmentally compatible, quick, and target oriented approach to improve nutrient use efficiency especially during critical growth stages or to remediate in-season deficiencies (Tariq et al. 2018). Although the amount of absorbed K via aerial tissues are much less compared to those absorbed through the roots, nevertheless, foliar applications of K are verified advantageous option over soil application (Ishfaq et al. 2016; Hussain et al. 2020). For instance, foliar spray of K during the latter growth stages improved potato yields up to 11% over soil application (Khan et al. 2010). Foliar application of KNO_3 or SOP may be economically viable option in K fertilization program. The foliar spray of K can also improve crop quality traits, i.e. the late-season foliar spray of K for cotton is beneficial during the heavy bollfilling stage for increasing lint yield and quality. Furthermore, it has distinguished advantageous under adverse soil conditions, for instance, salt stress (Khan and Aziz 2013) and drought stress (Ahmad et al. 2015; Hussain et al. 2016; Tariq et al. 2018).

8.3 Potash Under Salinity and Other Abiotic Stresses

Globally, abiotic stresses lead to dramatic impact on agricultural practices and production. It is predicted that major factors of abiotic stresses including salinity, heat stress, cold, drought have massive impact on worldwide agriculture, and they can largely diminish yield of major field crops (Wang et al. 2013). Plants exposed to stresses, namely high light intensity, drought, chilling, heat, and salinity suffer from oxidative impairment catalyzed by reactive oxygen species (ROS) (Cakmak 2005). Future climate changes are projected to be more challenging, for instance, higher in rainy summers and dry springs in northern Europe and extended dry periods in the southern Europe. Even it is revealed that abiotic stresses result in larger yield reduction in contrast to biotic stresses (Oerke 2006).

Adequate K availability can largely contribute in the resistance of nearly all biotic and abiotic stresses, in both direct and indirect ways. It is well documented that K supply can lead to restore photosynthetic rate, protein synthesis, enzymes activation, plant growth, and yield under various environmental stresses in several crop species. There are collective evidences from different research studies that adequate supply of K can lessen the unfavorable buildup of ROS by improving photosynthetic electron transport or reducing the membrane-bound NADPH oxidases. In contrast, K deficiency not only upregulates relevant K transporters, but also cause the synthesis of molecules such as ROS and the phytohormones, i.e. jasmonic acid (JA), auxin, and ethylene. Furthermore, adjustment of metabolic processes and upregulation of distinct transport proteins, K deprivation triggers developmental retorts in roots, all these processes allow plants to compete and survive under potash variable environments. Hence, K triggers numerous metabolic processes in plants, whereas its depletion result in increased vulnerability to environmental stresses.

8.3.1 Potassium and Salinity Stress

Detrimental effects of salt stress are well illustrated on growth and yields of various agricultural crops, particularly on irrigated land. The main consequences of its unfavorable impacts include osmotic effect which reduce the water availability to plants, higher salt stress lead to specific ion toxicity, and deficiencies of other plant nutrients, i.e. K and Ca. It inhibits seed germination, effect on leaf anatomy and plant physiology as well as photosynthesis, energy production, protein synthesis, and lipid metabolism. As roots are directly exposed to a saline environment, therefore reduce water uptake as a result of higher negative water potential in the rooting medium

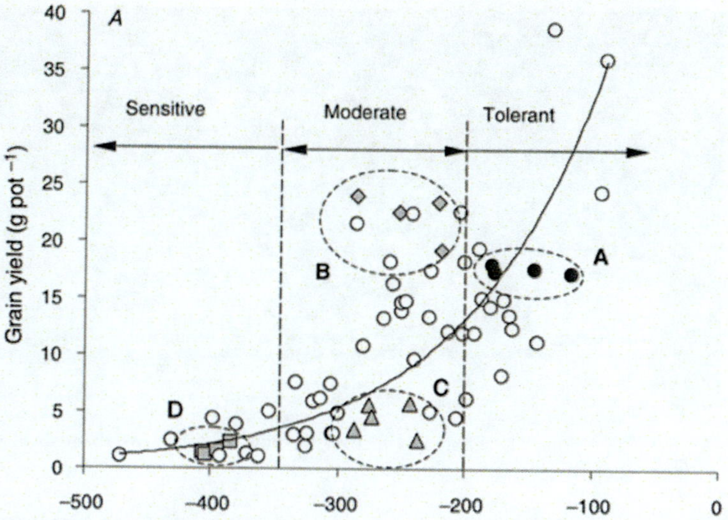

Fig. 8.1 A correlation between barley grain yield and NaCl-induced net K^+ flux measured from 3-day-old seedlings (adapted from Chen et al. 2007)

consequently lower acquisition and translocation of mineral nutrients. Furthermore, inhibited photosynthetic activity and closure of stomata mediated by lower K nutritional status encourages the generation of toxic ROS under salt stress (Cakmak 2005).

The interaction of Na^+ and K^+ is well recognized in plant nutrition which exposes that due to their similar charges, they can replace each other entirely in definite nonspecific metabolic functions in many plant species (Wakeel 2013). As Na^+ restricts the metabolic process therefore it cannot accumulate in the plant cytoplasm. The halophytic crop species not only respond to elevated Na^+ by regulating low cytosolic Na^+ level and high K^+/Na^+ ratio, in addition, improves plant growth which is largely subjected by Na^+ availability. In contrast, Na^+-sensitive plants are not proficient to exclude it at the root surface or accumulate in vacuoles. However, elevated K level can suppress the deleterious influence of salinity since the excessive Na^+ uptake into the plant cell is reduced with K^+ availability (Figs. 8.1 and 8.2).

The availability of Na^+ can substitute certain K^+ functions in plant, for instance, in sugar beet beneficial effects of Na^+ can be observed under limited K^+ supply (Fig. 8.2), nevertheless it also induced calcium (Ca^{2+}) deficiency (Wakeel et al. 2009). Meanwhile, K^+ supply reduces Na^+ stress, while Na^+ in order can partially relieve the K^+ deficiency up to some extent and promote plant growth in the appropriate amount (Rodríguez-Navarro and Rubio 2006; Fig. 8.3). However, the degree to which Na^+ to substitute K^+ varies between plant species, and even various cultivars of the same plant species. Additionally, enzyme activation is a quite distinct function of K^+, and Na^+ cannot substitute it in this context because of its larger hydrated radius and the transport of Na^+ within plant is restricted. Therefore,

Fig. 8.2 An elevated K supply can recover the adverse effect of salt stress (4 weeks old wheat seedling; photo by Abdul Wakeel)

Fig. 8.3 Similar shoot growth of sugar beet was experienced in the K applied treatment and the treatment where K was substituted by equivalent amount of Na (Wakeel 2008)

numerous K^+-specific enzymes cannot activate under K deficiency, and if Na^+ bind with them, then natural conformation of these enzyme can unbalance and binding of substrate to the enzyme can diminish (Mengel 2007).

In Pakistan, due to arid and semi-arid climatic conditions, the upward movement of salts in soil is higher predominantly in irrigated areas, which result in the development of more salt-affected soils. In interpretation of the diverse functions of K in Na resistance, an enhanced crop performance with K nutrition in salt-affected soils is frequently experienced (Wakeel 2013). Recently, it is reported that elevated K^+ nutrition under salt stress conditions mitigates the second phase of stress by reducing Na^+ toxicity in wheat shoot through decreasing water potential, and ultimately improves grain yield (Gul et al. 2019). The advantageous effects of potash supply under salt stress are also reported in other studies conducted under local conditions (Khan and Aziz 2013; Aown et al. 2012).

8.3.2 Potassium and Water Stress

Worldwide, crops are progressively being exposed to high temperature and drought stresses with detrimental impacts on agriculture productivity. The major limitation for optimum crop productivity in arid and semi-arid regions including Pakistan is inadequate accessibility of irrigation water. Crops that are unceasingly exposed to drought which inhibit photosynthesis activity results in misallocation of electrons which can generate ROS, finally leaf damage (Cakmak 2005) and, eventually, reduction in crop yield. A linear relationship between K supply and its influences in molecular and physiological mechanisms of plant under drought resistance is well established.

The adequate K nutrition during drought stress is crucial in phloem loading and translocation of sucrose to sink organs, hence improve root growth, surface area, and longevity. In contrast, inadequate K supply reduces root biomass and consequently hampered water and nutrient uptake including K per se. The exogeneous application of K is proposed advantageous under such conditions. Furthermore, K also improves water retention in plant tissues and regulates stomatal movement and, hence, controls plant water loss via transpiration (Fig. 8.4). However, maximum stomatal closure for the duration of the day time can restrict photosynthetic CO_2 fixation and may promote the generation of ROS. Subsequently, there are evidences that a main share of K nutrition in the mitigation of drought stress is not as a result of improved stomatal regulation, but because of non-stomatal influences on photosynthesis, phloem loading, primary metabolism, in addition to photoassimilates translocation towards sink organs (Cakmak 2005; Marschner 2012).

The adjustment of a satisfactory water status during drought stress is pivotal for plant survival, and K^+ mediates a vital role in osmotic adjustment with lowering osmotic potential, even under drought states (Marschner 2012). Under drought stress conditions, plant roots regulate water and nutrient uptake capability by adjusting K^+

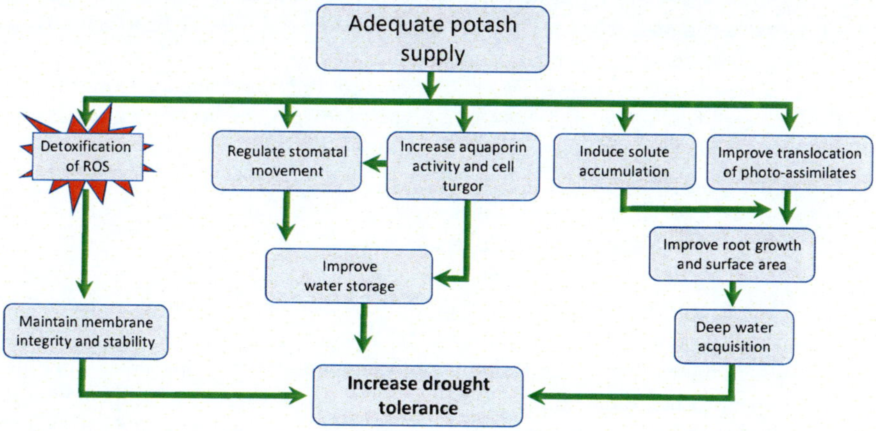

Fig. 8.4 Schematic illustration of adequate potash nutrition in mitigation of drought stress

channel and *PIPs* (plasma membrane intrinsic proteins) at the transcription level to adapt under water deficiency (Galmés et al. 2007; Cuéllar et al. 2010). For instance, in rice under drought stress, *PIPs* and K^+ channel-encoding genes were induced by K deficiency and downregulated by a polyethylene glycol (PEG)-mediated water deficit, which may lead to reduction in the membrane water permeability and followed by, promote cellular water conservation (Liu et al. 2006). As K^+ transporters/channel and water channels are literally co-regulated in plant osmoregulation to adopt cytosolic osmolarity and adjust them under drought or other stresses (Liu et al. 2006). Hence, an adequate K nutrition is indispensable for water relations, drought resistance, WUE and eventually crop growth (Fig. 8.4).

The climatic conditions of Pakistan are arid and semi-arid, and insufficient availability of irrigation water is main constrain in crop production. Under adverse climatic conditions including drought; the use of potash fertilizers can be promising to improve agricultural productivity. The evidences of K nutrition to mitigate adverse influences of drought stress under agroecosystems of Pakistan are growing. For instance, foliar application of potash increased grain yield and drought tolerance in wheat (Aown et al. 2012). In another study, adequate potash supply exhibited considerable improvement in yield and quality of sugar beet under water deficient conditions (Mubarak et al. 2016). Most recently, it is revealed that early maturing *Bt* cotton demands higher K nutrition under water stress to enlarge seed-cotton yield (Shahzad et al. 2019). However, the use of potash under drought stress or water shortage is often not appreciated by Pakistani farmers.

In addition to drought stress, waterlogging/flooding also has considerable consequences in yield reduction consistent with the crop species and plant growth stage, duration of the stress and soil type; accordingly, substantial economic consequences in many areas. Under waterlogging conditions, respiration of soil micro-organisms and roots reduces the residual oxygen, and the environment becomes hypoxic and, far along, anoxic. Additionally, as time of flooding expands, possibly toxic compounds, i.e. sulfides, soluble manganese and iron, CO_2, ethylene, ethanol, lactic acid, formic acid, and acetaldehyde, accumulate and affect the plant growth (Shabala 2011). The scientific literature supports that potash nutrition may cope with the undesirable effects of waterlogging up to certain extent. For instance, K supply under waterlogging conditions not only improved cotton growth, photosynthetic capacity, and photosynthetic pigments, in addition increased nutrient acquisition (Ashraf et al., Ashraf et al. 2011a, 2011b).

8.3.3 Potassium and Cold Stress

Cold stress hampers crop growth and development resulting in the reduction of overall productivity. Frost stress can be a critical yield limiting factor predominantly in regions with short vegetation periods, and numerous crops are susceptible to frost damage. Like other abiotic stresses, cold stress can also hamper photosynthetic activity and effectiveness of antioxidant enzymes, which lead to ROS generation damaging plant cells. It influences plant directly by hindering metabolic processes and indirectly by manipulating cold-induced oxidative, osmotic and other stresses. The role of potash in protecting plants against frost damage by lowering the cell solution freezing point has been characterized for many years. It has been revealed that only one night of chilling exposure under lower tissue K concentrations might results in enough crop damage to be comparable to the fertilizer expenditure of throughout season (Kant and Kafkafi 2002).

At low temperature stress, the activity of enzymatic processes and other transporters can suppress. Likewise, it also hinders K-dependent processes, i.e. photosynthetic CO_2 fixation, phloem loading, and metabolism, which result in the generation of ROS due to leakage of unutilized electrons. Meanwhile, the activity of enzymes which involves in ROS defense mechanisms improved in cold-sensitive crop, namely maize by seed treatment with KCl (Farooq et al. 2008). Hence, K supply can improve plant durability under cold stress by reducing ROS generation and improving antioxidant level. The elevated K concentration in the cell sap may lower its freezing point and ultimately reduce the deleterious impact of frost. Furthermore, the higher ratio of unsaturated to saturated fatty acids in phospholipid rich cell mediated by high potash supply can also justify to cope with frost stress by improving membrane fluidity. Additionally, modified cytosol K^+ level is indispensable for enzymes activities which are involve in frost resistance (Kant and Kafkafi 2002).

In brief, potash supply can hamper chilling damage and improve cold resistance by enhancing antioxidant level, reducing ROS production, lowering cell freezing point, and by increasing phospholipids. There are enough evidences that support potash nutrition role against cold stress, and frost damage is largely negatively correlated in numerous research studies with K supply (Cakmak 2005; Römheld and Kirkby 2010; Oosterhuis et al. 2013; Wang et al. 2013). In Pakistan, the impact of frost stress is perceived in numerous field crops (for example, sugarcane, tomato, chilies, guava, lemon, etc.) during the winter season, and K fertilization can be valuable to hamper yield reduction induced by chilling damage.

8.4 Potash to Reduce Crop Lodging

The displacement of plants from their vertical stance is termed "lodging," which can be reversible or irreversible depending on the degree of bending. It is subjected by numerous factors including adverse weather conditions, topography, soil type, disease, and management practices. It is commonly experienced in cereals such as wheat, corn, rice, oat, barley, as well as in sugarcane. Generally cereal lodged can be observed in the field at anthesis or early grain filling stage which result in both yield and quality reductions. Intensively cultivated crops with high inputs of irrigation and fertilizer application are more susceptible to lodging. However, adaptation of appropriate management practices such as time and method of planting, cultivation of lodging resistance cultivars, population maintenance, crop rotation, and disease management can alleviate risks of lodging. Furthermore, adequate crop nutrition,

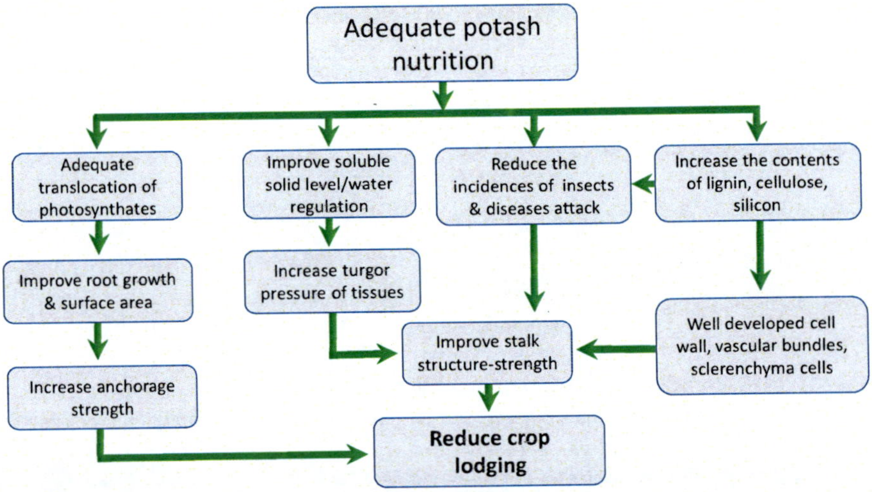

Fig. 8.5 Schematic illustration of adequate potash supply in mitigation of lodging incidents

especially the supply of N and K in adequate ratio can suppress lodging incidences through strengthening plant stem.

In most research reports the reduction in lodging incidents with K application are cited with special emphasis on its roles in root development, stalk structure strength, and carbohydrate metabolism (Fig. 8.5). When stem extension takes place, the maintenance of proper nutrition which mediates stem strength predominantly potash, manganese, and copper can reduce crop susceptibility to lodging. It appears likely that the stem strength, and probably the anchorage strength may improve with K fertilization. For instance, balanced application of K and N in rice improved basal internode space, cellulose content, stem-fiber, and silicon contents which resist lodging (Zaman et al. 2015). In contrast, elevated N without or limited K supply, excessive soil moisture, high seeding rates, and use of tall varieties can increase the risks of lodging. Being held in the cell vacuoles, K participates in water regulation in the plant, which maintain the plant strength. While under K deficient conditions, the movement of water into cells suppress, reduce turgor of plant tissues, and anchorage strength which ultimately increase the potential for lodging. For instance, due to poor root growth, lodging occurred at the base of rice under insufficient K supply (Bhiah et al. 2010).

At farmers' fields both root lodging and stalk breakage are reported which influences flowering, reduced photosynthetic CO_2 assimilation affecting the yield, ultimately. Potash fertilization promotes the energy status of plants and consequently the synthesis of numerous biochemical compounds such as lignin, cellulose, lipids, and vitamins. Therefore, lodging subsequent from poor fertility has frequently been attributed to a K deficiency as it encompasses in diameter of stem and stem strength. As poorly developed cuticles and thin cell walls under poor K nutrition enhance the crop susceptibility to diseases and pest attack which also trigger lodging.

Alternatively, increased N supply especially at early growth stages makes plant taller with the reduced amount of lignified tissue and root branching which eventually support lodging. Furthermore, under poor soil fertility status, the sole applications of N, P, or K does not improve the lodging resistance, therefore balance fertilization is crucial for optimum crop productivity.

In brief, a close relationship exists between the K nutrition and stem breaking strength, as K supply is principally allied with lignification of vascular bundles, sclerenchyma cells, and strengthening culms, in that way, improve lodging resistance. Furthermore, at adequate K nutrition plants produce photosynthate efficiently which utilize during grain development and inhibit to drop soluble solid level in stalk. Meanwhile, limited K supply reduces the diameter of stem which makes plant vulnerable to lodging. Although contribution of potash in crop lodging is largely accepted, so far farming practices have not been adopted in Pakistan with intention of K nutrition to mitigate lodging prevalence. Furthermore, there is little research efforts available under local conditions to support the relative consequences of plant nutrition to mitigate adverse effects of lodging.

8.5 Potash for Biotic Stresses

In the field, crops are challenged by numerous pathogens and insects/pests which triggers substantial yield losses. In addition to yield reduction, farmer price premium reduces from those crops which are infested with pests and/or diseases owing to higher production costs. The availability of mineral nutrients has a considerable influence on the tendency of crops to be infested by diseases or pests. Nevertheless, in contrast to human nutrition where the consequences of balance and enough diet is well acknowledged, the execution of balance nutrition for plant to enhance resistance and tolerance lags its potential. Although crop damage caused by pests cannot be completely avoided with sole nutrient management, but it can resist up to significant extent. Therefore, integrated pest management (IPM) diminish the risk of undesirable pesticide residues with diminishing the requirement for other chemical pest and disease control measures. Furthermore, farmer can lessen the input cost by adopting balanced fertilization with reducing the use of agrochemicals.

The ratio between N and K apparently participates a certain role in the host/pathogen relationship. It is broadly acknowledged that high potash supply in crops declines the incidence of pests and diseases while oversupply of N can influence in the opposite direction. The favorable effect of K has been supported by its intermediated role promoting the synthesis of high molecular weight compounds, for instance, starch, proteins, and cellulose, in that way reducing the level of low molecular weight compounds in plant tissues, i.e. organic acids, soluble sugars, amides, and amino acids (Fig. 8.6). Low molecular weight compounds usually persuade for the expansion of infections and insect invasions. Furthermore, adequate K fertilization can improve phenol concentrations in plant, which could also involve considerably in plant resistance against biotic stresses (Prasad et al. 2010).

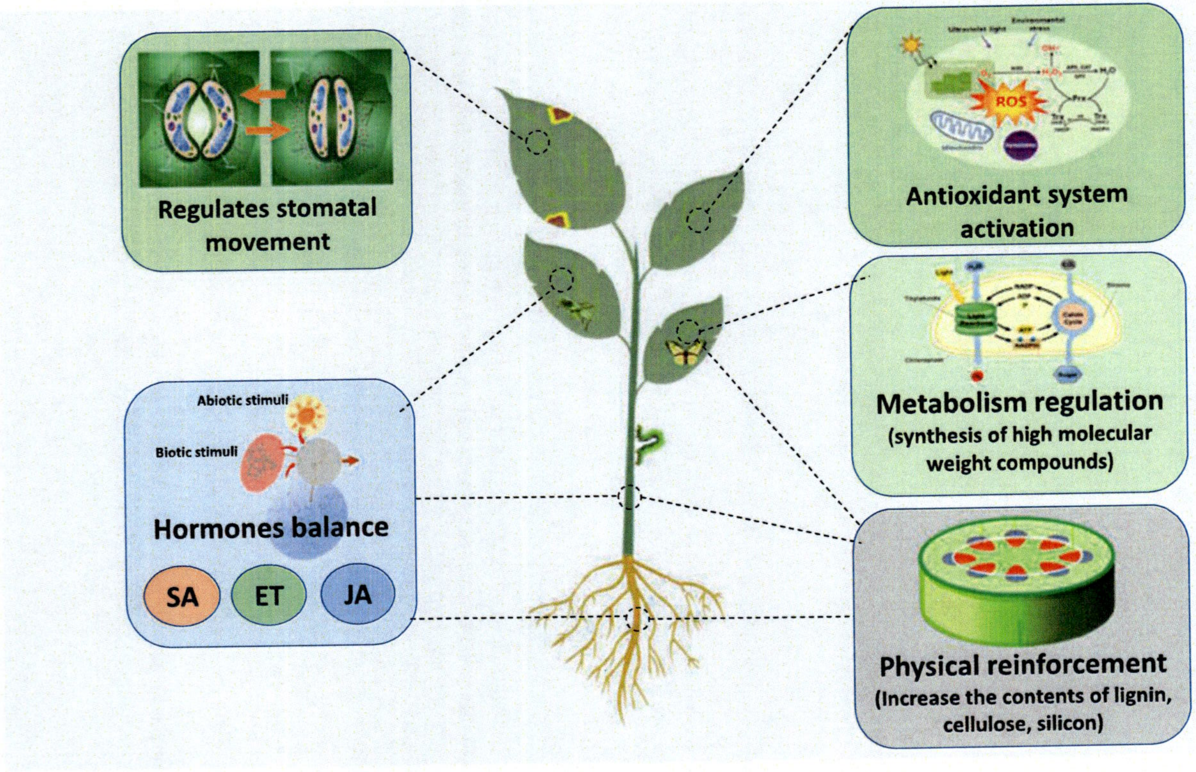

Fig. 8.6 Potassium protects plant from adverse effects of biotic stresses by improving structure and strength of cell wall, vascular bundles, sclerenchyma cells; synthesis of high molecular weight compounds, activation of antioxidant system, regulation stomatal movement, and hormone balance

Therefore, insects and pathogens are more predominant in K deficient crops which are higher susceptibie to pest attack and diseases (Marschner 2012).

The ample K nutrition produces a strong and vigorous plant, which can provide resistance against the insect and diseases damage more efficiently than those plants which are under stress. As mentioned earlier, under the limited K supply plant are more vulnerable to lodging, cracks, lesions and fissures which ultimately develop favorable microclimate with the extra humidity to fascinate pests. In contrast, plant physical resistance against pathogens can improve with sufficient K supply because it regulates stomatal movement and enhances the lignification of vascular tissue (Fig. 8.6). Number of research findings revealed the advantageous effect of an adequate K nutrition against pathogen damage. However, the effect of K on the incidence of pests and diseases varies depending on K nutritional status or forms and amounts of fertilizer applied or to both these factors (Amtmann et al. 2008).

The improved resistance to insects/pests with K application has been associated with numerous mechanisms, for instance, reduced cell permeability and susceptibility of tissue to insect attack. It has been observed in a scientific study that the rate of rice borer infestation was higher in treatment where no K was applied, meanwhile, it reduced swiftly at the elevated K level (Sarwar 2012). As the plants grown in K deficient soil are more vulnerable to infection than those plants which are grown under sufficient K supply (Wang et al. 2013). Additionally, the combined supply of N and P persuades the thicker cuticles and cell wall development, which acts as mechanical barriers to infection and invasion by pathogens, consequently insects/pests attack. In contrast, plant with ample supply of N and without considering sufficient K nutrition, the cell walls of plant stems can be thinner and soft tissue as a result of rapid growth with less resistance to chewing and sucking insects.

8.5.1 Potassium and Pathogens

Largely speaking, a contrary relationship is reported between K nutrition and the incidence of plant disease in most research studies (Table 8.3). Much earlier, it was revealed that K nutrition can reduce the incidences of fungal diseases approximately 70%, bacterial about 69%, viral around 41%, insect and mite infestation nearly 63% , and nematodes about 33% (Perrenoud 1990). As K supply mediates tissue cell structures, biochemical, and other metabolic functions consequently reduce vulnerability to pathogens. For instance, K fertilization observed effective in systemic

Table 8.3 Number of published papers reporting the consequences of potash nutrition on plant disease incidence (Prabhu et al. 2007)

Biotic factor	Decrease in disease	Increase in disease	No effect	Total
Fungi	89	33	8	130
Bacteria	18	5	0	23
Viruses	9	5	3	17
Nematode	3	6	1	10

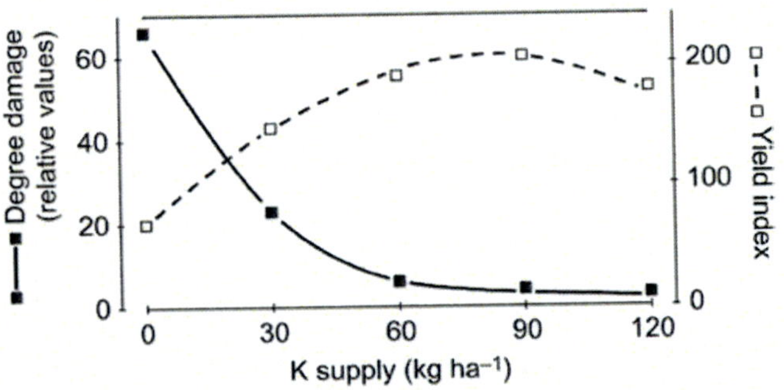

Fig. 8.7 Grain yield of wetland rice and incidence of stem rot (*Helminthosporium sigmoideum*) at different K supplies (Source: Marschner 2012)

protection in contradiction of powdery mildew. Leaf spot in cotton, stem and pod blight, and the purple seed stain in soybeans are correlated to low K concentration in soil, plant tissue or petiole. Plant vulnerability to infection under insufficient K supply particularly with fungal pathogens has been attributed to its function in phloem transport and primary metabolism (Römheld and Kirkby 2010). In potatoes, K supply has been reported to reduce the incidence of numerous diseases, for instance, dry rot, late blight, early blight, and powdery scab. As mentioned earlier phloem loading is restricted in potash deficient plant therefore, organic acids and soluble sugars increased in shoot tissue which attract pathogen for food.

Potash deficient plant diminishes the ability to assimilate N and aggravates chlorosis at the margins of their older leaves, i.e. cereals which increases vulnerability to diseases and affecting plant metabolism (Zörb et al. 2014; Coskun et al. 2017). Accumulated inorganic N in the sap of K-deficient plants via its restricted metabolism offers a more auspicious condition for bacterial growth and consequently prone to wilt. Furthermore, K deficient plants are imperfect in the regulation of stomata, thus resistance counter to bacterial pathogens is probable to be weak in these plants. Additionally, in contrast to vigorous plants, weak and succulent plants are observed to be more infested by pathogens. For instance, stem rot, *Helminthosporium sigmoideum*, usually arises at high N application in potash deficient soil while K fertilization can reduce their incidence and ultimately yields improvement (Fig. 8.7). The inverse relationship between K nutrition and disease incidence also revealed in other crop species such as the brown leaf spot in rice (*Helminthosporium oryzae*), and the rice blast (*Pyricularia oryzae*) or the sheath blight of rice (*Thanatephorus cucumeris*).

In Pakistan, there are no any notable efforts made regarding potash fertilization to mitigate adverse effects of pathogens. A significant knowledge gap exists for precise and applied efforts concerning potash fertilization to mitigate crop susceptibility to pathogens. Furthermore, the relationship of K nutrition and pathogen induced responses with special emphasis on transcriptomic intermediated signaling pathways may be valuable and new aspect on K-disease interactions.

8.6 Potash for Sustainable Agriculture

Agriculture is the mainstay to feed 222 million population of Pakistan and it contributes nearly 23% to country's total GDP. Crop production relies on climatic conditions and natural resources, and as dynamic system it demands excellent management for sustainable production. Concept of balanced fertilization comprises of not only improve crop productivity but also maximize profitability in the context of economic returns for every unit of fertilizer applied. Meanwhile, the imbalanced use of fertilizer is always uneconomical and discouraging for growers. Since green revolution potash is highly overlooked for crop production in Pakistan (see Chap. 10 for detail description) which is alarming for soil fertility and agriculture sustainability.

From preceding sections, it becomes clear that adequate K nutrition is crucial in biological functions to attain desirable crop yield and produce quality under certain optimal growth conditions. Potash fertilization is beneficial in arable areas to boost current stagnant crop yields. Thus, the use of K is indispensable for sustainable agricultural production; and its deficiency result in poor root growth, retarded growth, less resistance to stresses, delayed crop maturity, small seeds and ultimately reduce agriculture productivity. Furthermore, little research efforts are made in country in the context of the economics of potash fertilizers to improve net returns, which highlight that crops response to applied K is exceedingly variable because of diverse agroecological zones and clay minerology.

With the introduction of intensive agricultural production systems or as a result of imbalanced use of fertilizers; soil available K level is prone to deplete to fulfill the requirements of high yielding fertilizer responsive crop cultivars. Unfortunately, not much attention has been given to promote balanced use of fertilizer in Pakistan. Furthermore, future environmental conditions are predicted to be unfavorable, the quantity and quality of crop production are meaningfully exposed to number of biotic and abiotic stresses. Use of N fertilizers without K fertilization lead to reduce agricultural productivity and environment concerns in various growing areas of Pakistan. Balanced fertilization program in combination with potash nutrition can be promising for sustainable crop production, yield, and quality, as well as diminish the environmental risks. In brief, site-specific potash fertilization is indispensable to reduce huge yield gap between local progressive farmers and smallholder farmers. Application of potash and growing of K efficient genotypes in K deficient soil can be auspicious for sustainable agriculture productivity.

References

Ahmad Z, Waraich EW, Ahmad T, Ahmad R, Awan MI (2015) Yield responses of maize as influenced by supplemental foliar applied phosphorus under drought stress. International Journal of Food and Allied Sciences 1(2):45–55

Akhtar J, Shahzad A, Qureshi RH, Naseem A, Mahmood K (2000) Testing of wheat (*Triticum aestivum* L.) genotypes against salinity and waterlogging. Pak J Biol Sci 3:1134–1137

Akhtar J, Saqib ZA, Sarfraz M, Saleem I, Haq MA (2010) Evaluating salt tolerant cotton genotypes at different levels of NaCI stress in solution and soil culture. Pak J Bot 42:2857–2866

Amjad M, Akhtar J, Anwar-ul-Haq M, Imran S, Jacobsen SE (2014) Soil and foliar application of potassium enhances fruit yield and quality of tomato under salinity. Turk J Biol 38:208–218

Amtmann A, Troufflard S, Armengaud P (2008) The effect of potassium nutrition on pest and disease resistance in plants. Physiol Plant 133:682–691

Annadurai K, Palaniappan P, Masilamani P, Kavimani R (2000) Split application of potassium on rice – a review. Agric Rev 21:36–44

Aown M, Raza S, Saleem MF, Anjum SA, Khaliq T, Wahid MA (2012) Foliar application of potassium under water deficit conditions improved the growth and yield of wheat (*Triticum aestivum* L.). Journal of Animal and Plant Sciences 22(2):431–437

Ashraf M, Afzal M, Ahmad R, Ali S, Shahzad SM, Aziz A, Ali L (2011a) Growth and yield components of wheat genotypes as influenced by potassium and farm yard manure on a saline sodic soil. Soil Environ 30:115–121

Ashraf MA, Ahmad MSA, Ashraf M, Al-Qurainy F, Ashraf MY (2011b) Alleviation of waterlogging stress in upland cotton (Gossypium hirsutum L.) by exogenous application of potassium in soil and as a foliar spray. Crop and Pasture Science 62(1):25–38

Azeem M, Ahmad R (2011) Foliar application of some essential minerals on tomato (*Lycopersicon esculentum*) plant growth under two different salinity regimes. Pak J Bot 43:1513–1520

Bhiah, K.M., Guppy, C., Lockwood, P. and Jessop, R. 2010. Effect of potassium on rice lodging under high nitrogen nutrition. 19th World Congress of Soil Science, Soil Solutions for a Changing World

Cakmak I (2005) The role of potassium in alleviating detrimental effects of abiotic stresses in plants. J Plant Nutr Soil Sci 168(4):521–530

Chen Z, Zhou M, Newman IA, Mendham NJ, Zhang G, Shabala S (2007) Potassium and sodium relations in salinised barley tissues as a basis of differential salt tolerance. Funct Plant Biol 34(2):150–162

Coskun D, Britto DT, Shi W, Kronzucker HJ (2017) Nitrogen transformations in modern agriculture and the role of biological nitrification inhibition. Nat Plants 3(6):17074

Cuéllar T, Pascaud F, Verdeil JL, Torregrosa L, Adam-Blondon AF, Thibaud JB, Sentenac H, Gaillard I (2010) A grapevine shaker inward K^+ channel activated by the calcineurin B-like calcium sensor 1–protein kinase CIPK23 network is expressed in grape berries under drought stress conditions. Plant J 61(1):58–69

Farooq M, Aziz T, Cheema ZA, Hussain M, Khaliq A (2008) Activation of antioxidant system by KCl improves the chilling tolerance in hybrid maize. J Agron Crop Sci 194(6):438–448

Galmés J, Pou A, Alsina MM, Tomas M, Medrano H, Flexas J (2007) Aquaporin expression in response to different water stress intensities and recovery in Richter-110 (Vitis sp.): relationship with ecophysiological status. Planta 226(3):671–681

Gul M, Wakeel A, Steffens D, Lindberg S (2019) Potassium-induced decrease in cytosolic Na+ alleviates deleterious effects of salt stress on wheat (*Triticum aestivum* L.). Plant Biol 21(5):825–831

Hussain M, Tariq AF, Nawaz A, Nawaz M, Sattar A, Ul-Allah S, Wakeel A (2020) Efficacy of fertilizing method for different potash sources in cotton (*Gossypium hirsutum* L.) nutrition under arid climatic conditions. PLoS One 15(1):e0228335

Hussain RA, Ahmad R, Nawaz F, Ashraf MY, Waraich EA (2016) Foliar NK application mitigates drought effects in sunflower (*Helianthus annuus* L.). Acta Physiol Plantarum 38(4) https://doi.org/10.1007/s11738-016-2104-z

Hussain Z, Khattak RA, Irshad M, Aneji AE (2013) Ameliorative effect of potassium sulphate on the growth and chemical composition of wheat (*Triticum aestivum* L.) in salt affected soils. J of Soil Sci Plant Nutr 13:401–415

Hussain Z, Khattak RA, Irshad M, Mahmood Q (2014) Sugar beet (*Beta vulgaris* L.) response to diammonium phosphate and potassium sulphate under saline-sodic conditions. Soil Use Manag 30:320–327

Ishfaq M, Mubarak MU, Cheema SA, Wakeel A (2016) Potassium uptake by maize roots as affected by foliar application of potassium. In: Conference paper: in: international conference on "significance of potash use in Pakistani agriculture" at: IAGS. Lahore, Pakistan, UOP

Kant S, Kafkafi U (2002) Potassium and abiotic stresses in plants. In: Pasricha NS, Bansal SK (eds) Potassium for sustainable crop production. Potash Institute of India, Gurgaon, pp 233–251

Khan A, Aziz M (2013) Influence of foliar application of potassium on wheat (*Triticum aestivum* L) under saline conditions. Sci Tech Dev 32(4):285–289

Khan AA, Khan MN, Inamullah SS, Arshad I, Muhammad I, Zeb A (2015) Effect of potash application on growth, yield and yield components of spring maize hybrids. Pure App Biol 4: 195–203

Khan MZ, Akhtar E, Safdar MN, Mahmood MM, Ahmad S, Ahmed N (2010) Effect of source and level of potash on yield and quality of potato tubers. Pakistan J Botany 42(5):3137–3145

Khan R, Gurmani AR, Gurmani AH, Zia MS (2007) Effect of potassium application on crop yields under wheat- rice system. Sarhad J Agric 23:277–280

Liu, H.Y., Sun, W.N., Su, W.A. and Tang, Z.C, 2006. Co-regulation of water channels and potassium channels in rice. Physiol Plantarum, 128(1), 58-69.

Majid SA, Asghar R, Murtaza G (2007) Potassium – calcium interrelationship linked to drought tolerance in wheat (*Triticum aestivum* L.). Pakistan J Botany 39:1609–1621

Marschner P (2012) Marschner's mineral nutrition of higher plants, vol 89. Academic Press, Oxford

Mengel K (2007) Potassium. In: Barker AV, Pilbeam DJ (eds) Handbook of plant nutrition. Taylor & Francis, Boca Ratan, FL, pp 91–120

Mubarak MU, Zahir M, Gul M, Farooq M, Wakeel A (2016) Sugar beet response to potassium fertilizer under water sufficient and water deficient conditions. e-ifc 46:30–32

Oerke EC (2006) Crop losses to pests. J Agric Sci 144(1):31–43

Oosterhuis DM, Loka DA, Raper TB (2013) Potassium and stress alleviation: physiological functions and management of cotton. J Plant Nutr Soil Sci 176(3):331–343

Perrenoud S (1990) Potassium and plant health. International Potash Institute, Basel, Switzerland

Pervaz H, Makhdum MI, Ashraf M, Shabab-ud-Din (2006) Influence of potassium nutrition on leaf area index in cotton (*Gossypium hirsutum* L.) under an arid environment. Pakistan J Botany 38: 1085–1092

Prabhu AS, Fageria NK, Huber DM (2007) Potassium nutrition and plant diseases. In: Datnoff LE, Elmer WH, Huber DM (eds) Mineral nutrition and plant disease. American Phytopathological Society, Saint Paul, MN, pp 57–78

Prasad D, Singh R, Singh A (2010) Management of sheath blight of rice with integrated nutrients. Indian Phytopathology 63(1):11

Rodríguez-Navarro A, Rubio F (2006) High-affinity potassium and sodium transport systems in plants. J Exp Botany 57(5):1149–1160

Römheld V, Kirkby EA (2010) Research on potassium in agriculture: needs and prospects. Plant Soil 335(1–2):155–180

Saifullah, Ranjha AM, Yaseen M, Akhtar ME (2002) Response of wheat to potassium fertilization under field condition. Pakistan Journal of Agriculture Science 39:269–272

Saqib M, Akhtar J, Qureshi RH (2008) Sodicity intensifies the effect of salinity on grain yield and yield components of wheat. J Plant Nutr 31:689–701

Sarwar M (2012) Effects of potassium fertilization on population build up of rice stem borers (lepidopteron pests) and rice (*Oryza sativa* L.) yield. J Cereals Oilseeds 3(1):6–9

Shabala S (2011) Physiological and cellular aspects of phytotoxicity tolerance in plants: the role of membrane transporters and implications for crop breeding for waterlogging tolerance. New Phytol 190(2):289–298

Shahzad AN, Rizwan M, Asghar MG, Qureshi MK, Bukhari SAH, Kiran A, Wakeel A (2019) Early maturing Bt cotton requires more potassium fertilizer under water deficiency to augment seed-cotton yield but not lint quality. Sci Rep 9(1):1–10

Soomro AF, Tunio S, Keerio MI, Rajper I, Chachar Q, Arain MY (2014) Effect of inorganic NPK fertilizers under different proportions on growth, yield and juice quality of sugarcane (*Saccharum officinarum* L.). Pure Appl Biol 3:10–18

Tariq M, Afzal MN, Muhammad D, Ahmad S, Shahzad AN, Kiran A, Wakeel A (2018) Relationship of tissue potassium content with yield and fiber quality components of Bt cotton as influenced by potassium application methods. Field Crops Res 229:37–43

Wakeel A, Ishfaq M (2016) Promoting precise and balanced use of fertilizers in Pakistan at farm-gate level. e-ifc 47:30–35

Wakeel A (2013) Potassium–sodium interactions in soil and plant under saline-sodic conditions. J Plant Nutr Soil Sci 176(3):344–354

Wakeel A, Abd-El-Motagally F, Steffens D, Schubert S (2009) Sodium-induced calcium deficiency in sugar beet during substitution of potassium by sodium. J Plant Nutr Soil Sci 172(2):254–260

Wakeel A. 2008. Substitution of potassium by sodium in sugar beet (*Beta vulgaris* L.) nutrition with special reference to K-fixing soils. Doctoral Degree Thesis. Faculty of Agricultural and Nutritional Sciences, Justus Liebig University Giessen, Germany

Wang M, Zheng Q, Shen Q, Guo S (2013) The critical role of potassium in plant stress response. Int J Mol Sci 14(4):7370–7390

Zaman U, Ahmad Z, Farooq M, Saeed S, Ahmad M, Wakeel A (2015) Potassium fertilization may improve stem strength and yield of basmati rice grown on nitrogen-fertilized soils. Pak J Agric Sci 52(2):439–445

Zörb C, Senbayram M, Peiter E (2014) Potassium in agriculture – status and perspectives. J Plant Physiol 171(9):656–669

9 Use of Potash in Pakistan

The use of chemical fertilizers was started in Pakistan after green revolution and gradually the use of nitrogenous and phosphatic fertilizers has been improved to a large extent. However, the use of potash is still very low making it more worsening for crop production due to imbalance in nutrients in the soil. Here we describe the chronological use of potassic fertilizers in comparison to other chemical fertilizers with critical analysis to identify the reasons of low K use in Pakistan specifically and in developing countries generally. Furthermore, the importance of K fertilizer use in Pakistan has also been highlighted.

9.1 Green Revolution and Status of Potash Use

Historically, introducing high yielding and fertilizer responsive cultivars in the 1960s, the usage of chemical fertilizer was promoted in many parts of the world including Pakistan. In the start, the usage of nitrogenous was endorsed, and phosphatic fertilizers were introduced later in the 1970s, and their import is expanding over time. However, the application of nitrogenous fertilizers was considerably higher with the possibility of quick, better, and more cost-effective response and accordingly, presently, nitrogenous fertilizers contribute 73–78% of total fertilizer consumption in Pakistan. Meanwhile, in accordance with last six decades record (1960–2017) of fertilizer consumption, potash has long been overlooked in country for profitable crop production (FAOSTAT 2019). The nutrients distribution of past decades illustrates that K shared nearly 1% of total nutrient consumption for agricultural use and the bias in the use of K is clear in Pakistan (Fig. 9.1).

After the green revolution, the higher production of food, fiber, and other crops also results in a much higher removal of nutrients from soils. The removal of K in comparison to N is higher in different agroecosystems of the country, principally vegetables and cereals lead to higher mining of croplands. Largely, crop requirements for K are equal to N and 3–5 times higher than P. Nevertheless, the

A. Wakeel, M. Ishfaq, *Potash Use and Dynamics in Agriculture*,
https://doi.org/10.1007/978-981-16-6883-8_9

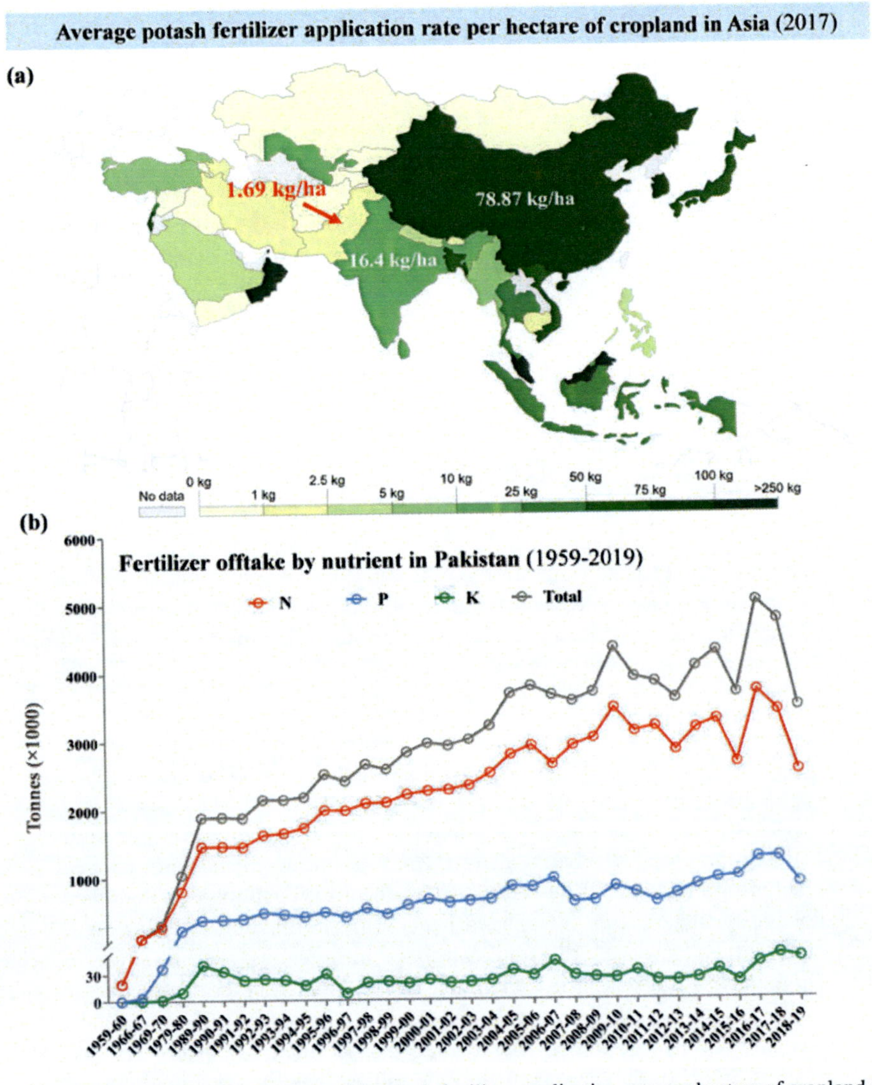

Fig. 9.1 Use of potash in Pakistan. (**a**) Potash fertilizer application rate per hectare of cropland in Asia, (**b**) historical trend of nutrient-wise fertilizer consumption in Pakistan since green revolution (Data source: FAOSTAT & National Fertilizer Development Centre (NFDC), Islamabad)

regular K consumption in Pakistan since the green revolution has been much less than N and P (Mian and Ahmad 2007). The slow and steady increment in phosphatic fertilizer demand continued, and since last two decades (2002–2017) the consumption of P fertilizers improved nearly 5% in Pakistan (Fig. 9.1). However, concern of such neglected or imbalanced K usage is evident in the altering K fertility status of soils in various cropping systems of Pakistan.

Table 9.1 Nutrient-wise fertilizer use (kg ha^{-1}) in distinct provinces of Pakistan

Province	Nitrogen	Phosphorus	Potassium	Total	N:K	P:K
Punjab	132.2	32.5	1.4	166.1	94.2	23.2
Sindh	156.2	42.7	1.4	200.3	111.5	30.5
KPK	111.4	21.8	0.9	134.1	123.8	24.2
Balochistan	88.7	14.1	1.0	103.8	88.7	14.1
Pakistan	132.4	32.4	1.4	166.1	94.5	23.1

Source: National Fertilizer Development Centre (NFDC), Islamabad

Furthermore, critical analysis of fertilizer application with crop yield indicates that instead of increment in N and P fertilizer consumption since last decades, the average crop production is stagnant. The limited use of K fertilizer may be one of the reasonable factors for such stagnant agriculture productivity of country, because for optimum, economical, and sustainable crop production balanced fertilization is indispensable. With the increase in nitrogenous fertilizer application, the N-K ratio has expanded considerably in various agroecosystems of Pakistan, which contributing to less economic returns and low NUE (Zhang et al. 2015; Shahzad et al. 2019).

According to the most recent report, nearly all soils of Pakistan are N deficient mainly due to arid to semi-arid climate resulting rapid decomposition of organic matter. Almost 90% soils are deficient to marginal in P, and approximately 50–60% are stated K deficient (Ahmad 2008). In Pakistan, the majority of farming community relies upon N and P fertilizers as per their awareness and economic resources, whereas the use of K is severely overlooked, for instance, nearly 44% cotton growing areas of Pakistan are K deficient while 16% farmers are using potash fertilizer (Wakeel and Ishfaq 2016). According to National Fertilizer Development Centre (NFDC), the average potash application per hectare is <1.5 kg throughout the country, in contrast to 132 kg for N and 32 kg for P fertilizers, which is literally alarming for the agriculture productivity and sustainability (Table 9.1).

Although Pakistani soils comprise of mica which contain about 7% K but as a result of continuously cultivation of high yielding N responsive varieties and almost no addition of potash is plausibility of crops harvest took away mass K from soil available fractions. Based on the annual crop production, Pakistani soils are predicted to mining 100–150 K_2O kg ha^{-1} $year^{-1}$. Soil samples analyzed by soil fertility research institute also exposed the apparent reduction in available K in distinct cropping zones of Pakistan (see Chap. 2). Furthermore, in most parts of the country, soil analysis is hardly used, and instead farmers are given a blanket recommendation which does not meet the crops' demand in many cases.

In Pakistan, crop residues and farm-yard manure (as manure cakes) are usually used for heating and cooking purpose without recycling potash rich ash or sludge back to arable land which obtains very low input of potash. As the residues of legume and cereals, i.e. roots, stalks, tops, and leaves of sugar, vegetable, and tuber crop comprises of approximately more than half the dry matter. Such bulk amount of crop residues which contains essential plant nutrients are removed and utilized them for other purposes such as cooking, animal feed, or biofuels, it means a considerable share of the nutrient is lost at the end of crop harvesting from cultivated areas. Consequently, a continuing decline in soil fertility status, together with soil organic matter and K contents can be predicted as a substantial factor in limiting agriculture productivity.

Various attempts have been made by local and international fertilizer organizations, nonetheless, potash fertilization for crop production has sustained miserably low. The imbalanced use of fertilizer is always uneconomical, and very expanded ratio for N to K consequences into numerous concerns, for instance, crop lodging mainly in wheat and rice. Even though, with the availability of compound fertilizers, for example, NPK has somewhat improved K use in the field crops, but then again circumstances remain consistent due to the limited availability SOP in the market. Even so, in last few years, various private companies introduced K containing liquid fertilizers which are suggested as foliar spray or fertigation, however, their adaptability by farming community is not visible. Besides, completely relying on these liquid fertilizers which contain little amount of active ingredient cannot accomplish the crop requirements.

The importance of maintaining an adequate supply of K for plants in order to support the high yield and quality of harvestable product is clear. The impact of K deficiency is frequently observed and accordingly, considerable K deficiency is projected in the coming decades with more soil depletion if imbalanced fertilization continuously carried out and even soil fertility can considerably decline. Hence, active management of K is critical for achieving optimal agricultural productivity.

9.2 Crop-Wise Fertilizer Use

Currently, the USA and China use nearly 40% of total global potash containing fertilizers, and other developing countries, for instance, India and few Asian countries are projected to considerably increase their K demand soon. Worldwide, cereals, especially maize, wheat, and rice have the maximum share (37%) in potash consumption followed by fruits and vegetables (22%), oilseeds (16%), cotton and sugar (11%), and others crop species (14%) (FAO 2011).

Pakistan has two distinct cropping seasons, “Kharif” including rice, sugarcane, cotton, maize, millet, sorghum crops, literally start from April to June and crop seems ready to harvest during October to December. Second is "Rabi," including wheat, gram, lentil, barley, tobacco, rapeseed, and mustard crops, sowing starts in October–December and harvested in April–May. The agriculture productivity of Pakistan considerably relies on the timely availability of irrigation water. In Pakistan, wheat is a staple and most fertilized crop which shares nearly 50% of total fertilizer consumption in the country because of staple crop and more inevitability of economic returns, therefore cultivated at huge area. After wheat, cotton is the second most fertilized crop and shares about 25% of total fertilizer consumption, followed by sugarcane (8%) and rice (6%). The other crops, fruits and vegetables are

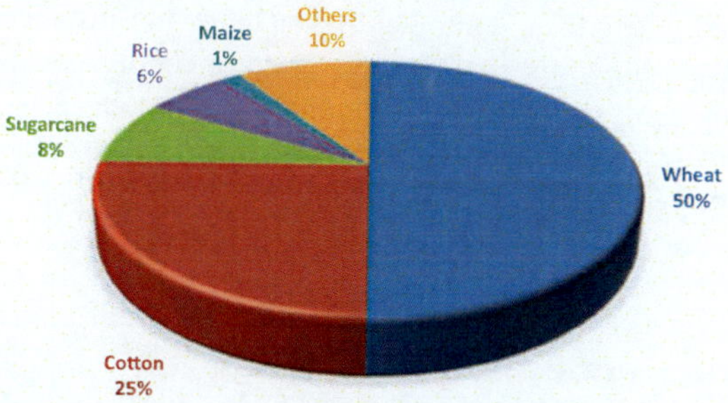

Crop	Nutrients tonnes (×000)
Wheat	2307.2
Cotton	1153.6
Sugarcane	369.2
Rice	276.9
Maize	69.2
Others	438.8
Total	4396

Fig. 9.2 Crop-wise usage of fertilizer (2018–2019) in Pakistan (Data source: Pakistan Bureau of Pakistan)

usually not fertilized like cereals and shares 10% of total fertilizer consumption in the country (Fig.9.2).

In Pakistan, farmers tend to prefer nitrogenous fertilizer (predominantly urea) due to their comparatively less price, quick response, and widespread availability. Though, we cannot achieve target yield with sole application of nitrogenous fertilizer, in addition it can deplete soils fertility status if applied continuously without addition of other nutrients. Therefore, a precise and balanced use of NPK and other essential plant nutrients must be considered if soils are to be managed in a sustainable manner. In Pakistan, potash fertilization has been depressingly low almost in all the major crops, except potato where farmers apply it for better yields and tuber quality.

The use of crop residues, farm-yard manure (FYM), green manures, and bio-fertilizers are negligible in Pakistan in the context of plant requirements. Furthermore, canal water is deliberated as a good source of potash and could moderately accomplish K requirements; nevertheless, the availability of canal water is scarce in major agricultural areas of Pakistan now days. For example, nearly 9% of the farming community in cotton growing areas of Pakistan totally dependent on canal water while remaining irrigating crops with tube-well water or mixing it with canal water in different proportions (Wakeel and Ishfaq 2016). Therefore, K status of soil is changing rapidly due to continuous depletion from soil via multiple cropping systems without its adequate supply, while its recommendation is still negligible compared to N and P in the agricultural soils of Pakistan.

Crop response to applied K depends on numerous factors, for instance, crops and varieties, soils, climatic conditions, and management practices. Crop yield levels and K demand of crop determine the response to applied K. Therefore, such uneven crop response to potash application shakes the confidence of farmers to K fertilization. Furthermore, there is a general misconception while recommending potash fertilizers in Pakistan that the calculation of appropriate K requirements is just soil testing with chemical extraction methods. This method, nevertheless, do not consider the consequence of K limitation with spatial availability as a significance of various root characteristics. As root architecture varies largely between crop types particularly between monocots and dicots, and even different genotypes of same crop which may differ obviously in root density, length, and number of root hairs. For instance, under drought stress K deficiency symptoms frequently appear in dicotyledonous and legumes crops even if soil analysis data shows enough indigenous K contents, therefore deliberating the significance of root growth and weather data can be beneficial to rationalize precise K availability for plant.

9.3 Recommendations of Potash for Crops

Nutrient uptake depends on an interaction of plant with soil environment, for instance, K uptake will be restricted if there is insufficient soil moisture for root activity. An overabundance of antagonistic cations, i.e. calcium (Ca) or magnesium (Mg) in the soil may impair K uptake through their competitive interaction. Many of the currently used soil-test based K fertilizer recommendations are largely developed in country for agronomic crops. However, the heavy demand of some horticultural crops during peak uptake periods is not being accounted with traditional soil testing techniques. Parameters other than total yield may also be influenced by an abundant K supply. For example, increasing K fertilization can improve the number of highly profitable, large and extra-large tomato fruit (Zhu et al. 2017). Furthermore, adequate supplies of K do not always immediately result in the loss of quality or yield.

Fertilizer recommendations are generally formulated in accordance with crop yield goals, soil fertility status, and sometimes farmer capacity. The recommendation of fertilizer for the high yield goal, site-specific knowledge about soil fertility status along with crop requirements is indispensable. Potash fertilizers are not always effective for all soils, and therefore, before the application of K containing fertilizers, it is essential to assess the amount of K in the soils which is to be fertilized and the degree of availability to plants. Soil minerology, crop type, management practices, and environmental conditions prominently mediate the availability of each form of K in the soil. To increase the efficiency of potash, and to get desire plant response, the understanding of dynamic reaction between different forms of soil K must be considered. It can be valuable in formulating the sound K fertilizer recommendation for soils and crop that is based on soil type rather than their blanket application.

The nature and amount of clay influences the rate of K release from non-exchangeable fractions in soil. As mentioned earlier soils containing vermiculite and montmorillonite have more K than soils containing Kaolinite. The native K status may subject to sources as well as the degree of weathering of parent materials. It is also important to note that K release depends on the structural compositions of minerals. As per stated in Chap. 5, soil extraction methods are largely used as the basis for potash recommendations primarily for exchangeable K (EK). However, they are fairly effective for various soils, but soils containing 2:1 type clay mineral demand satisfactory calibration (Mengel and Kirkby 2001). Generally, soils that have exchangeable K of <80 mg kg^{-1} soil are considered to be fertilized with 100 kg K_2O ha^{-1}. The recommended amount of fertilizers in Pakistan are in the ratio of 2:1:0.5 as NPK, respectively (Table 9.2).

Table 9.2 Fertilizer recommendations provided by Punjab Agriculture Department for major crops of Pakistan (kg ha^{-1})

Crop	Nitrogen	Phosphorus	Potassium
Cotton	200–400	88–175	95–125
Wheat	80–130	58–115	62
Rice	85–172	68–102	62–80
Maize	250	115–172	62–125
Potato	250	125	125
Sugarcane	172–300	58–172	62–125
Sugar beet	125	125	150
Oilseed crops	32–88	32–88	32–62
Sunflower	150	100	65
Kharif fodders	58	58	32

Source: http://dai.agripunjab.gov.pk/croptechnologies

The soil nutritional status mediated by plant roots is largely regulated by root physiology and morphology of different plant species. Usually, grasses have more extended root system than dicots. Therefore, the rate of potash acquisition per unit root length is lower in grasses as compared to dicots and, consequently, dicots comparatively respond earlier to K fertilization. There are general observations that sugar beet is more efficient, compared to potatoes and wheat, in the extraction of slowly exchangeable K (Steingrobe and Claassen 2000). Furthermore, K efficient potato cultivars seem to mobilize slowly exchangeable K most efficiently by means of secreted root exudates in contrast to inefficient cultivars (Trehan and Sharma 2002; Trehan et al. 2005). Therefore, the role of crop cultivar in the variation of nutrient uptake cannot be overlooked (Ishfaq et al. 2018). Additionally, characterization of interlayer K dynamics is important in calculation of K uptake via crops, especially soils containing 2:1 type clay mineral, but unfortunately it is underestimated in Pakistan due to the lack of availability of any routine approach for its determination.

The 4R Nutrient Stewardship principles are same globally (right source, right rate, right time, and right place), but how they are manifested locally varies depending on site-specific characteristics. In the case of K applications, 4R represents a way of thinking that can help growers set goals for enhancing crop quality and profitability, reducing K loss, maintaining soil K fertility, and promoting site-specific K management. Many crops cultivated in different agroecosystems of Pakistan respond to potash fertilization considerably (see Chap. 8). Usually, tuber and root crops have high demand of K. Largely, crop requirements for potash cultivated on medium K soils are as per: fruit crops > vegetables > pulses and oil seeds > cereals. However, understanding and knowledge of K dynamics in soil evidences to be the best method for precise fertilizer recommendation for crops. Therefore, much remain open to elucidate in this direction that may be valuable in the rationalization of the long-term K releasing capacity of Pakistani soils for future adequate recommendations to farmers.

9.4 Causes of Low Potassium Fertilizer Use

Generally, it is supposed that due to limited financial resources, smallholder Pakistani farmers are unable to use potash as an agricultural input. Despite this, ground reality may differ, as same farmers are applying nitrogenous and phosphatic fertilizers, very expensive pesticides, and bearing others input cost. The lack of financial ability can be one of reason for low acceptability of potash containing fertilizers in Pakistan, but not the sole reason. Some other constrains such as farmers' perceptions and awareness, capacity of agriculture extension staff, government policies, and instability of market prices are likely responsible. Doubtlessly, below mentioned facts can be leading constraints of low popularity of potash fertilization among Pakistani farming community.

9.4.1 Lack of Awareness About Balanced Fertilizer Use Concept

Majority of uneducated farmers in Pakistan along with such massive migration of educated youth to cities have expanded a gap in the context of acceptability of new technology. A huge number of farming community still relies on traditional sources to get agriculture information, for instance, family, imitate neighbors or "veteran" farmers, and have weak perceptions of evidence-based practices. Poor informal education and lack of awareness regarding the consequences of imbalanced use of fertilizers is one of the reasons for low acceptance of potash fertilizers among local growers. There is not any notable farmers' awareness campaign available in Pakistan to pursue significance of balanced fertilization, in contradiction of our neighbor country China (Cui et al. 2018). A reasonable number of farmers in Pakistan deliberate that application of higher amount of fertilizer (principally urea) guarantees higher yield rather than bearing in mind that distinct nutrient has its own consequences.

Slow crop response to potash as compared to N and P may be realistic reason for its insignificant acceptance. As soil response to potash fertilization is dissimilar and very diverse than others (Wakeel et al. 2013), therefore, farmers' awareness in the framework of potash nutrition and its distinct consequences on crop physiology and productivity are required. A satisfactory approach to motivate growers towards balanced use of fertilizer including K to mitigate the adverse effects of abiotic stresses in crops (Cakmak 2005). Furthermore, participatory method can be beneficial to improve potash use, as the effectiveness of this tactic was noticeable in the very high increment in the use of nitrogenous fertilizers after 1960s. Likewise, the acceptability of diammounium phosphate (DAP) enlarged among farming community due to its positive and better crop response than other phosphatic fertilizers or probably for its better response may be due to its N gradient contribution.

9.4.2 Canal Irrigation Water Supplies Sufficient Potash: A Common Misconception

A very common misconception prevails among the local farming community that K fertilizer is not needed in the soils which are irrigated with canal water. Though canal water is a good source of potash, but it is not enough to accomplish the needs of high yielding crop species. In earlier times, when Pakistani soils were not as exhausted as they are now and crop cultivars were comparatively low yielding, this contribution may have been adequate. But now, on the one hand, soil fertility status is depleted as a result of continually cultivation of high yielding and fertilizer responsive crop varieties without organic or inorganic inputs of potash. On the other hand, extended climatic changes and silting of water reservoirs; reduced the adequate availability of canal water to irrigate all soils and therefore, a substantial share of irrigation water is supplemented via underground water (Wakeel and Ishfaq 2016).

In brief, K gradient in canal water can be complementary to soil applied potash, predominantly in canal water irrigated areas, but cannot satisfy the crop nutritional prerequisite to attain target yields. Furthermore, with enhancing the use of ground water which carries very small fraction of potash will further decrease its contribution from irrigation water, subsequently, alarming threat in the reduction of soil fertility status of the country.

9.4.3 General Crops Fertilizers Recommendations

In Pakistan, fertilizer recommendations are generally specified by the provincial agriculture department, according to the overall soil fertility status of whole province/region. Site-specific or crop response-based recommendations are very exceptional in Pakistan. Because of insufficient soil testing facilities or awareness, farmers normally do not deliberate their soil fertility status, and the wide ranging or conventional trend of fertilization are practiced. Although, few soil and water testing laboratories have been established at district level, but then again, they may not be enough in number to entertain all farmers, or smallholder farming community are not much aware to attain precise recommendations.

The blanket fertilizer recommendations may be reasonable for N and P, but for K, they often do not bring about the satisfactory crop response because soil is highly heterogeneous in terms of K dynamics and mineralogy. This may be one of the probable reasons for the acceptability of low potash fertilizers among growers and thus, it is difficult to motivate farmers to apply an additional amount of fertilizers. Consequently, farmer perceptions uphold that potash is not required for their soils, and it is difficult to convince them if they are not obtaining any considerable advantage with additional fertilizer input in the term of crop yield and economic return. Therefore, comprehensive approaches are needed to create awareness and to update farming practices (see Chap. 10). As potash fertilizer recommendation should be always in line with site-specific K status of soil.

9.4.4 Fertilizer Prices and Availability of K Fertilizer

With small land holdings (~2.5 hectares) and inadequate financial ability are limitations for farmers to adopt balanced use of fertilizer. The high costs of K fertilizers may hesitate farmers when they already buying expensive N and P fertilizers. Although certain literature in Pakistan highlight the consequences of potash application for sustainable and economic crop production, nonetheless increasing prices of potash fertilizers with lack of financing is a considerable limitation for the balanced use of fertilizer. Less attention of government and no aggressive awareness campaign by agrotechnicians or fertilizer industries to promote potash use have embarrassed the condition. Furthermore, another cause for insufficient potash usage in the country is that limited fertilizer companies are involved to deliver potash and accordingly had limited accessibility of potash in the stocks of local supplier.

Several research findings are expanding in Pakistan highlighting the significance of K application, which confirm an obvious crop response to K fertilization in terms of crop yield and quality grown under diverse conditions (see Chap. 8). Nevertheless, there is no noticeable policy has been implemented by the government to instate research-based findings for the agriculture productivity of the country. Additionally, fertilizer industries are also ambiguous on the matter of concern, which result in rare adoptability of potash fertilizers in Pakistan. Subsidizing agriculture inputs particularly fertilizers, counting potash can be valuable to uplift agriculture productivity and sustainability of the country. The utmost example is the promotion of N containing fertilizers when urea was subsidized after 1960s (green revolution), later on farmers adopted it due to its substantial impact on crop yield. Such substantial approaches may be advantageous to highlight the consequence of K and to promote its adoptability at ground level.

9.4.5 Instability in Market Prices of Agricultural Produce

Marketing standards of agricultural commodities and satisfactory price premium are key incentives for farmers to adopt balanced use of fertilizer and other finance-based new technologies. The prices of farming inputs and products are unpredictable which generates insecurity in economic returns and reduce the farmers' motivation to optimize agricultural inputs. For instance, instability in the prices of cotton and rice are creating frustration as a result of farmer loss regardless of massive expenditures for buying of inputs. Furthermore, there is no as such prominent infrastructure of quality-based prices of agricultural products, for instance, sugar industry purchases sugarcane from farmers on the weight basis rather than based on sugar recovery (sucrose content). Therefore, farmers remain afraid of low-price production with high usage of inputs. Under such instability of price premium, the adoption of high yielding agricultural practices reduced especially among smallholder farmers.

References

Ahmad N (2008) Fertilizer use practices in Pakistan. In: Hamid A, Jalil A (eds) Balanced fertilizer use: impact on crop production. Proc Symp National Fertilizer Development Center, Islamabad, Pakistan, pp 65–71

Cakmak I (2005) The role of potassium in alleviating detrimental effects of abiotic stresses in plants. J Plant Nutr Soil Sci 168:521–530

Cui Z, Zhang H, Chen X, Zhang C, Ma W, Huang C, Zhang W, Mi G, Miao Y, Li X, Gao Q (2018) Pursuing sustainable productivity with millions of smallholder farmers. Nature 555(7696):363–366

FAO (2011) Current world fertilizer trends and outlook to 2015. Food and Agriculture Organization of the United Nations, Rome

FAOSTAT (2019) U.N. food and agriculture organization, corporate statistical database.

Ishfaq M, Kiran A, Khaliq A, Cheema SA, Alaraidh IA, Hirotsu N, Wakeel A (2018) Zinc biofortified wheat cultivar lessens grain cadmium accumulation under cadmium contaminated conditions. Int J Agric Biol 20(12):2842–2846

Mengel K, Kirkby EA (2001) Principles of plant nutrition, 5th edn. Kluwer Academic Publishers, Dordrecht, p 849

Mian SM, Ahmad Z (2007) Potassium scenario in Pakistani Punjab. In Proc international symposium on balanced fertilization for sustainable crop productivity. Punjab Agri Univ Ludhiana, India,vol II, p 428–429

Shahzad AN, Qureshi MK, Wakeel A, Misselbrook T (2019) Crop production in Pakistan and low nitrogen use efficiencies. Nat Sustain 2(12):1106–1114

Steingrobe B, Claassen N (2000) Potassium dynamics in the rhizosphere and K efficiency of crops. J Plant Nutr Soil Sci 163:101–106

Trehan SP, Sharma RC (2002) Potassium uptake efficiency of young plants of three potato cultivars as related to root and shoot parameters. Commun Soil Sci Plant Anal 33:1813–1823

Trehan SP, El DH, Classen N (2005) Potassium efficiency of 10 potato cultivars as related to their capacity to use non-exchangeable soil K by chemical mobilization. Commun Soil Sci Plant Anal 36:1809–1822

Wakeel A, Ishfaq M (2016) Promoting precise and balanced use of fertilizers in Pakistan at farmgate level. e-ifc 47:30–35

Wakeel A, Gul M, Sanaullah M (2013) Potassium dynamics in three alluvial soils differing in clay contents. Emirates J Food Agric 25:39–44

Zhang X, Davidson EA, Mauzerall DL, Searchinger TD, Dumas P, Shen Y (2015) Managing nitrogen for sustainable development. Nature 528:51–59

Zhu Q, Ozores-Hampton M, Li Y, Morgan K, Liu G, Mylavarapu RS (2017) Responses of tomato to potassium rates in a calcareous soil. HortScience 52(5):764–769

10 Strategies to Improve Potash Use in Pakistan

As described in the previous chapters, the use of potash in Pakistani agriculture is very low and a strategic move is required to enhance its use for agricultural sustainability and quality food production. Starting from misconceptions and farmer awareness fertilizer subsidies have been recommended through arguments.

10.1 Elimination of Misconceptions

The present fertilizer application approach in Pakistan is not balanced and has led to widespread deficiency of various mineral nutrients in soil. Potassium, being a major nutrient is severely overlooked in the fertilizer management strategies. Intensive crop cultivation is continuously being experienced in most of the agroecological zones of Pakistan, and therefore, the issues of nutrient depletion are certain. The degradation of soils, imbalanced use of fertilizers, continuous crop removal, inefficient use and management of nutrient, and irrigation water led to K deficiency in Pakistan. The mining consequences of intensive cropping systems are oriented on the production of succeeding crops. Thus, to maintain optimum crop production, it is essential to understand soil fertility status, crop nutrient requirement, and the amount of nutrient removed from the soil through crops.

Canal water is considered as a valuable source of potash and may fulfill crop K requirements to some extent, however, its contribution further decreased due to insufficient supply in major cultivated areas of Pakistan. Due to intensive agricultural practices, it was projected that nearly 0.265 million Mt of K are depleting annually from the arable lands of Pakistan (Bajwa 1994), whereas the annual addition of potash fertilizers is about 0.03 Mt K_2O. More interesting study in the Punjab province indicate and annual decrease of ~2 mg kg^{-1} K_2O from the cultivated soils, arguing that current practices for soil fertility management are not adequate to sustain the soil productivity. This situation is very uncertain for agriculture sustainability in Pakistan. For that reason, there is an instantaneous need to

A. Wakeel, M. Ishfaq, *Potash Use and Dynamics in Agriculture*,
https://doi.org/10.1007/978-981-16-6883-8_10

eliminate misconception regarding use of potash fertilizer for sustainable and economical crop production in Pakistan.

Potassium has become a limiting element in multiple cropping systems, where inadequate potash supply has result in the depletion of indigenous soil K reserves. Crop production with imbalanced fertilization is always uneconomical and discouraging for farmers, with the expanded ratio of N:K lead to certain undesirable concerns, for example, crop lodging grown in certain areas. Furthermore, the amount of K present in the soil solution at any point of time represents only a very small amount of the total soil K which is not enough for the crop requirement in a growing season. High total K concentration does not guarantee for sufficient supply to crops. Potash is appearing a limiting nutrient in distinct agroecosystems of Pakistan, for example, report regarding the available soil K in Punjab province which has the core agricultural soil contributing around 70% to total agricultural production of the country, display that it reduced to the average annual rate of 3 mg kg^{-1} in the last 15 years (Mian et al. 2009). This reveal that total depletion of approximately 60 mg K kg^{-1} soil from Punjab province during two decades, therefore the soils which were considered to have sufficient indigenous K contents earlier are now approaching the deficiency threshold.

Depletion of indigenous soil K reserves as a result of multiple cropping systems has become a matter of concern. In Pakistan, the field-specific potash application and incorporation of crop residues in the soil during farming practices can be appreciated to improve soil K fertility and ultimately crop yield. Moreover, the K application for optimum profit production can be regulated on a crop-by-crop basis, for example, maize response to applied K is extremely high and uneven for soils in Pakistan. Therefore, a plant-based approach for field-specific nutrient management is needed for developing efficient K management. In brief, K is being removed from fields due to the introduction of high yielding crops and intensive agriculture, therefore K fertilizers are required for agricultural sustainability. In the agricultural scenario of Pakistan, the following benefits of potash fertilization have been experienced:

1. The application of potash to K deficient soils improves crop growth and yield.
2. Potassium improves the quality of agricultural commodities with increasing protein, oil, and vitamin C content in cereals. In fruits and tuber crops, K increases the size and improves color, flavor, storage, and shipping quality.
3. Potassium improves resistance against drought, salinity, frost, and pests.
4. Synergistic effect of K to N and P improves the efficiency of nitrogenous and phosphatic fertilizers.

10.2 Creating Awareness Among Farmers

It seems to be a potential challenge to improve the dissemination of scientific based knowledge on a ground level for farming practice. There is literally a huge gap between scientific research, and its dissemination and successful application in the farmer field. The agricultural research is meaningless if it does not serve farming

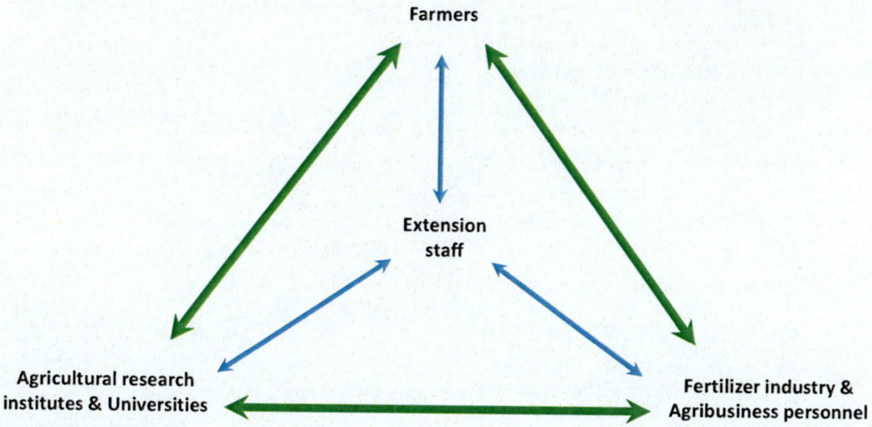

Fig. 10.1 Bilateral and close interactions are needed between scientists, fertilizer industry, farmers along with extension service to promote precise and balanced use of fertilizers with special reference to potash use in Pakistan

community. To accomplish the challenge a closer and bilateral cooperation among agriculture scientists, agrochemical industry, extension staff, and farmers is indispensable (Fig. 10.1). Therefore, the immediate future challenge is to ensure appropriate scientific based knowledge to circulate for the benefit of farming community. Understanding the soil nutrient status and corrective fertilizer management practices to support high yields of high-quality crops requires a continuation of scientific information and broadly-based education programs for the farm advisory service, soil and plant testing laboratories, trainers, the mass media, and ultimately the farmer.

Research, its application, and ultimately dissemination of any technology is mediated by certain factors, and it is difficult to achieve it single-handedly. Many factors such as education level, training, capacity of agrotechnicians and extension staff, financing, and farmer access to information media affects to adopt it. As K fertilization is one of the key incomes earning inputs which enable farmers to produce lower unit cost crops. Balanced fertilization and reduction of K mining can maintain soil fertility which will prevent smallholder farmers from falling into poverty and ultimately improve their livelihood. Therefore, a great challenge exists to promote the use of potash in farmers' fields according to integrated soil–crop system management. The methods include number of technology interventions and training approaches which increase the effectiveness of information dissemination, for instance, jointly with extension service farmer gathering, crop seminar, result demonstration, group discussions, soil testing service, print and electronic media, etc. (Fig. 10.2).

Growers should attain certain levels of training that are directly related with their farming practices and should be committed to acquire new skills to update their conventional practices. Furthermore, trustworthy bodies and well-organized mass

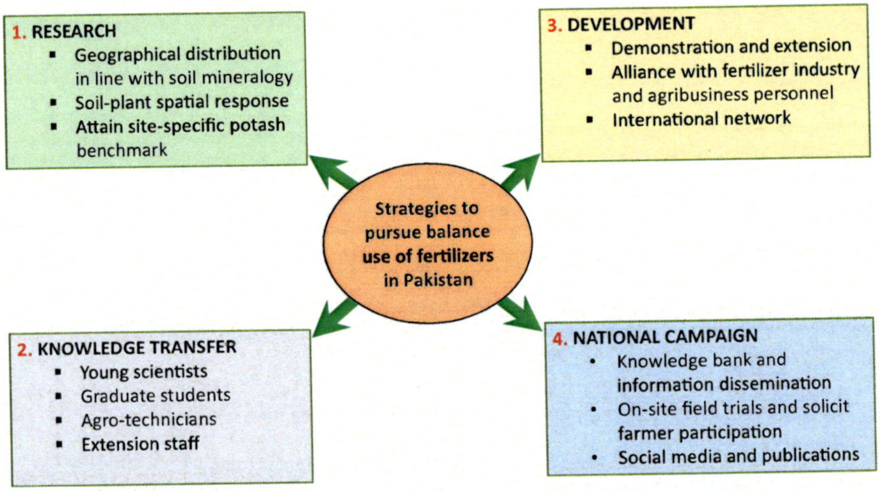

Fig. 10.2 Schematic strategies to pursue sustainable productivity with millions of smallholder farmers through promoting precise and balanced use fertilizes in Pakistan

media can perform an imperative role through advertisement in order to stimulate the adoption new technology among farming community. Additionally, extension staff or other agrotechnicians should update their knowledge connecting with rationalized technology that will be delivered to farmers. Concisely, scientifically verified and evidence-based farming practices must be adopted by the farmers to improve local agriculture productivity.

Furthermore, changing behavior demands building trust, which takes efforts and time. For instance, when initiate to promote precise and balanced use of fertilizers, farmers often assume whether trying to sell them something, i.e. pesticides or fertilizers. Once solid trust develops with higher yield and less fertilizer application, factually in the fields of progressive farmers, and with no any hidden agenda, then farmers can become willing to update their farming practices.

The establishment of a Potash Research Institute in Pakistan with intentions of deepen research, innovate nutrition regulation ways, technology, products, and technical service. It will support to carry out knowledge transfer under local conditions with the combination of plant nutrition basis and applied research as efficiently working in our neighbor countries, for example "Potash Research Institute of India" or "International Magnesium Institute" in China.

10.3 Subsidizing Potash Fertilizers

One of the main reasons for imbalanced uses of fertilizers in Pakistan is price disparity in case of N contrasted to P and K. Potash containing fertilizer lags far behind in this race as a result of price remain high and continue to rise, which is projected to have undesirable impact on crop yields, quality of agricultural commodities, and sustainability of soil productivity. Several policy approaches can be adopted to promote balanced use of fertilizer among the smallholder farming community. Government can take the initiative in the context subsidizing on target-oriented fertilizers, facilitating farmers to purchase fertilizer by providing credit, monitoring responsible institutes, import tariffs, procurement and distribution, and regulation of retail markets to reduce farm-gate prices.

The repeated fluctuations in fertilizer policies and provision of a dual fertilizer market such as subsidized and free market can minimize the targeted distribution of subsidized fertilizer. Therefore, regardless of frequent changes in fertilizer policy scenarios over the years, support for the potash fertilizer price subsidy should largely be constant. Furthermore, to ensure the most effective and profitable use of imported K fertilizer, priority should be accorded to making K fertilizers available in the distinct regions or districts where soil indigenous K contents are low or crops with high K requirement are grown. Grant targeted subsidies of K fertilizers to smallholder farmers that have limited resources to purchase fertilizer on a commercial basis with the establishment of regional fertilizer procurement and distribution services can be valuable.

The other often-observed challenge in Pakistan is an inadequate supply of potash fertilizers. In addition to adequate supply, fertilizer quality monitoring and regulatory roles can be supportive to realize crop response to applied potash fertilizer among farming community. The collected fertilizer samples from the local retail market or any other agro-dealer markets can be tested in the laboratory on regular interval to ensure the quality of available fertilizers. Furthermore, training of manpower for fertilizer analysis in the country is also obligatory. Additionally, to develop grower trust, the sale of fake, misbranded, adulterated, in addition to underweight fertilizer bags can be duly prosecuted and punished.

The initiation of voucher programs and distribution among farming community can be market-smart form of subsidy to ensure the most effective K fertilization. They will enable small holder farmers to purchase a specific quantity of K fertilizer from approved dealers who agree to receive vouchers as payment. They may be used by either reducing the price of fertilizer or providing farmers with repayment expected at some later date. The dealers can then exchange collected vouchers from voucher program organizers, or government, regularly with an approved margin to secure their expenses and profit. Furthermore, vouchers system can be practiced targeting specific groups of farmers, for instance, in potash deficient agricultural areas or farmer cultivating crops with high K requirement. They can develop additional demand for fertilizer and accordingly improve fertilizer market expansion if the pursued farmers are not already using fertilizer.

Fertilizer subsidies can vary in terms of (a) point: namely farmer, trader, domestic fertilizer producer; (b) form: for example, voucher, cash payment, reduced market price, transport subsidy; and (c) direct (fertilizer price reduced), or indirect, for instance, reducing taxation on fertilizer. Furthermore, initiating incentives for fertilizer importers or manufacturers for the promotion of target fertilizers or reducing fertilizer distribution costs such as by improving road infrastructure to decrease high transport costs can be advantageous. In brief, prices of agricultural commodities are very uncertain in country; therefore, the government should provide subsidies on agricultural inputs, including potassic fertilizers. To promote precise and balanced use of fertilizers in Pakistan following points can be appreciated:

1. Make sure self-reliance in the country fertilizer requirement by imported and/or local production
2. Subsidize expensive fertilizers, i.e. K fertilizer to make it affordable for smallholder farmers
3. Make certain that the good quality of fertilizer is attainable to farming community at the right time in the right area.

10.4 Developing Economic Potassium Recommendations

The basic goals of fertilizer management are to make certain for applying the right source of nutrient, at the right rate, at the right time, and in the right place, and the beneficial effects verified through improve soil health, better crop performance, reduced environmental concerns, and ultimately improvement of ecosystem services and net economic return. To achieve these goals, well planned efforts are needed at national levels to justify soil total potash, K-fixing capacity of the soils, exchangeable and non-exchangeable K under different soil–crop-climatic conditions to optimize K recommendations in country.

In Pakistan, there is great potential to ensure balanced use of fertilizer and well-organized potash fertilization in relation to other nutrients particularly N, Mg, and certain micronutrients. Furthermore, determination of soil exchangeable K and correlating crop response to applied fertilizer in term of yield can be extensively and successfully adopted on various soils. Meanwhile, the exchangeable K extraction method has verified not good enough tool for K recommendations specifically for soils comprising of 2:1 clay mineral as they can release as well as fix K at interlayer sites. Additionally, determination of K requirements for crop through chemical extraction method may be unsatisfactory due to the certain limitation of potash acquisition by plant roots which can be mediated by soil physical and chemical factors. For instance, soil spatial clay mineralogy, water stress conditions, low pH, salinity or compaction along with plant factors such as crop species and genotype, root distribution within the soil profile can limit the K availability.

It is of primitive importance to keep an inspection on depletion by systematic monitoring to certify that K is not limiting factor in optimum crop production and to apply potash with appropriate ratio of other mineral nutrients to avoid its deficiency. Crops, particularly cereals with well branched root system, draw large share of potash from non-exchangeable pool. Furthermore, in soils low in available K, cultivation of K efficient genotypes can enhance its resource use efficiency and accordingly more net economic return (Zia-ul-hassan et al. 2014). Furthermore, K and N teamwork promote to achieve maximum economic yield rather than sole N fertilization (see Chap. 11).

In brief, only soil analysis data might unsatisfactory to make the best conclusion of K fertilizer recommendation. Soil test information is certainly valuable, nevertheless deliberating other factors, for example, crop demand dynamics, timing of application, and source of potash fertilizer are all have great consequence. Therefore, soil K status alone does not always ensure that there will be reasonable response to applied potash. Besides, foliar application of K during high demand, for instance, fruit development or under transient drought-induced K deficiency can be valuable to improve net economic return. In low-fertility areas, it needs to justify what can be economic returns with increasing fertilizer use? Furthermore, potash fertilizer use in certain areas may contribute partially to these goals owing to unsatisfactory soil fertility status, unpredictable calamities, and/or market circumstances make fertilizer unprofitable. Thus, fertilizer promotion programs can ensure cost-effective contributions of recommended fertilizer in terms of expand the productivity and profitability under a wide range of smallholder farmer conditions.

Long-term site-specific field trials with adjusted potash fertilizers can rationalize its economic recommendations for distinct areas. Furthermore, GIS-based data counting long-term weather forecasting and soil mineralogy for distinct agroecosystem is needed to optimize K fertilizer requirements in the country. The application of potash particularly in the rainfed cultivated area may be high because of reduced release or its availability from soil. Furthermore, in irrigated areas, K gradient participating from canal water can be measured and it can be considered with additional potash recommendation (Ahmad et al. 2014). This may be valuable to optimize potash recommendation as well as its economical application. It is well documented that incorporation of organic amendments, for instance, compost, farmyard manure, vermicompost, etc. along with chemical fertilizers can improve the status of all the fractions of K. Manure application is positively correlated to soil health, crop yields and net economic return (Du et al. 2020). Additionally, crop response to applied N during the year of N application, while the response of K application on many soils is experienced in the subsequent years due to lasting effects of K (Syers et al. 2008).

10.5 Export Quality Agricultural Productions

Agriculture is a risky business and crops may not persist satisfactory as a result of several biotic and/or abiotic stresses. Therefore, enterprises and government institutions should play a significant role in developing strategies for marketing of local agricultural commodities in international markets. Strategies for the production and promotion of quality agricultural production can be achieved with bilateral cooperation. There are numerous contributors involved in efficient and profitable agriculture, and local agricultural commodities can be introduced in international markets if all participants cooperate lucratively.

The use of potash and its positive impact both on crop yield and quality is well documented. But unpredictable prices of agricultural commodities, lack of quality premium on cash crops, for instance, cotton, sugarcane, fruits, and vegetables, and higher prices of potash fertilizers often discourage local farmers. Potash improves the quality of agricultural products, and to obtain desirable quality produce we may have to go beyond the general K recommendations for getting higher yield. Therefore, to improve the quality of agricultural produce, especially for high value export agricultural commodities, there is a need to apply extra K fertilization. The application of potash improves appearance, taste, and flavor as well as the shelf life of fruits and vegetables which promotes export of agricultural commodities and decreases the risk of rejection when the product is offered for sale (see Chap. 6).

Generally, Pakistani smallholder farmers are too poor, too weak, and too isolated to sale their products in a reasonable price; therefore, they cannot earn any significant price premium. Negligible support from government and enterprises is also constraint for farmers to consider the quality of their farm product. Quality based agricultural product premiums always remain high in international markets because consumers increasingly require high-quality products and are willing to pay more for them. Therefore, developing infrastructure to export the quality agricultural product will motivate the farmers to improve the quality through the application of K fertilizers and ultimately earning reasonable price premium.

10.6 Quality-Based Prices for Agricultural Commodities in Local Markets

The revenue of agricultural commodities received from the markets is largely mediated by two basic principles: the first is the quantity of product transported to the market and the second is the quality of the product offered to the market for sale. Though product quantity is considered the leading element for the revenue generation, however, quality traits of the agriculture products can also regulate considerably in terms of price premium. Indeed, for certain crops, especially for fruits and vegetables quality performs key or dominant part in generating net revenue or sometimes bonus price. If farmers can attain higher and stable margins of their produce, then they will encourage to adopt and/or improve their farming practices including balanced use of fertilizer.

Adequate K supply can facilitate in quality development of numerous crops. Additionally, plant grown with sufficient potash supplied express lesser incidences of diseases, pests, cracks, lesions on vegetables and fruits which improve taste and appealing from consumer side (see Chap. 6). Meanwhile, when the appearance and quality of product is poor, the farmer cannot sell their produce for the best possible price. Furthermore, K application is also important for the nutritional point of view, for instance, content of protein, oil, vitamins, etc. Many field trials conducted by IPI have revealed that balanced fertilization with potash facilitates the farmer to produce agricultural commodities which meets the various quality criteria. Hence, the marketing system of Pakistani agricultural commodities is needed to be modified in term of quality, particularly in case of cash crops. Furthermore, there is prerequisite to minimize margin received by middle men to facilitate farmers to obtain fair price of their product.

To promote K fertilization in Pakistan, the prices for agricultural commodities need to be based on the quality. The presence of suitable macroeconomic policy, along with supportive marketing and pricing policies of agricultural commodities can ensure resourceful market operation and competition. If few industries have usual monopolies, they can be duly prosecuted. Agro-based industries, for example, sugar industry should participate per se in the research and development to improve the quality of input and to introduce sucrose content-based sugarcane purchasing. As K application to sugar crops such as sugarcane and sugar beet increases the sucrose content and ultimately sugar recovery (see Chap. 6).

10.7 Role of International Organization

Worldwide, there are number of international organizations established with mission to promote precise and balanced use of fertilizer, including use of potash, to reduce further decline of soil fertility and ensure food security along with more nutritious food. However, a reasonable constrains that always exist under Pakistani agricultural scenario are: how effectively can mobilize smallholder farmers, what target investments are required to improve overall farming capacity, and how can international organization can collaborate locally and effectively promote the use of balance fertilization on the wide range.

The common objectives of social programs are poverty mitigation or improving the livelihood of rural areas. Promotion of balanced and effective use of fertilizer by international organizations, generally making up all or a considerable part of fertilizer imports. Additionally, the general role of international organization can be to stimulate farmers to optimize their farming practices that reduce the costs of agricultural production and ultimately improve their living standard. The greater collaboration with local institutes for research and development, capacity-building, for instance, farmer training, workshops, expansion of farmer organizations and easing of rural financial markets to promote smallholder farmer capacity to purchase fertilizer can be valuable. There are many research and development projects effectively accomplished or under consideration in Pakistan with collaboration of international organizations, for example, United States Agency for International Development (USAID) and International Plant Nutrition Institute (IPNI). Additionally, International Potash Institute (IPI) promotes balanced use of fertilizers in number of countries, including Pakistan. IPI presently publishes in 25 languages including Urdu (national language of Pakistan), which facilitates farmers in number of countries to access information regarding the significance of balanced fertilization.

For instance, in Pakistan, two consecutive years internship programs were inaugurated by International Potash Institute (IPI) in selective regions with intentions to highlight the consequences of precise and balanced use of fertilizers directly to farmers at their farm-gate level (Wakeel 2015; Wakeel and Ishfaq 2016). More than 1400 farmers holding nearly 23,740 ha of cultivated land were covered by selected interns, and internship reports have been published on the IPI website:

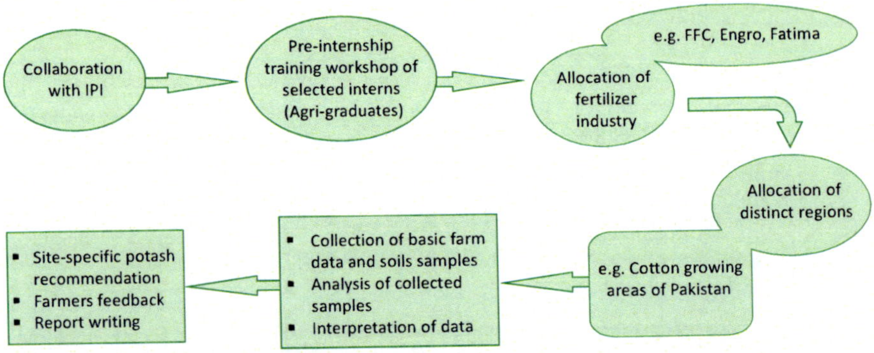

Fig. 10.3 Schematic illustration of approach adopted to promote precise and balanced use of fertilizers in Pakistan at farm-gate level with collaboration of International Potash Institute (IPI), Switzerland and local fertilizer industry (2015–2016)

https://www.ipipotash.org/publications/publication-453 and https://www.ipipotash.org/publications/eifc-405. These internship programs were supported by all local stakeholders, fertilizer companies, progressive farmers, extension staff, and such direct dissemination of information to farming community via agriculture graduates found effective approach to emphasize the significance of potash in Pakistan agriculture (Fig. 10.3). Such initial steps can be base line to elucidate and promote site-specific potash fertilization in distinct cropping systems of Pakistan considering soil mineralogy and crop spatial response.

References

Ahmad S, Khan MZ, Akhtar ME, Zubair M, Farooq MA, Afghan S (2014) Nutritional assessment of sugarcane on the basis of irrigation water quality and soil tests. Pak Sugar J 1:12–20

Bajwa MI (1994) Soil potassium status, potassium fertilizer usage and recommendations in Pakistan. Potash Rev 20(3):128

Du Y, Cuia B, Zhanga Q, Wanga Z, Suna J, Niu W (2020) Effects of manure fertilizer on crop yield and soil properties in China: a meta-analysis. Catena 193:104617

Mian SM, Ali I, Shaheen K (2009) Annual progress report. Soil Fertility Research Institute, Punjab, Lahore Pakistan

Syers JK, Johnson AE, Curtin D (2008) Efficiency of soil and fertilizer phosphorus use: reconciling changing concepts of soil phosphorus behaviour with agronomic information. FAO fertilizer and plant nutrition bulletin 18. Food and agriculture Organization of the United Nations, Rome, 109.

Wakeel A (2015) Balanced use of fertilizers in Pakistan: status and perspectives. International Potash Institute, Switzerland

Wakeel A, Ishfaq M (2016) Promoting precise and balanced use of fertilizers in Pakistan at farm-gate level. e-ifc 47:30–35

Zia-ul-hassan MA, Basra SMA, Rajpar I, Shah AN, Galani S (2014) Response of potassium-use-efficient cotton genotypes to soil applied potassium. Int J Agric Biol 16(4)

Future Research Perspectives

11

Based on critical discussion in the previous chapters, future research aspects have been described here to achieve the targets recommended for sustainable crop production in Pakistan. This is a crux of the book for researchers to move forward for research related to potash in developing countries.

11.1 Precise Potassium Recommendations

It is well recognized that field specific and crop season specific fertilizer application can significantly improve their uptake, nutrient use efficiencies, and ultimately high economic return. Site-specific nutrient management (SSNM) is an approach of nutrient management which covers contribution from soils, required from fertilizers to formulate fertilizer recommendation and requirement for unit yield. The SSNM demands intensive soil sampling and analyses so as to develop soil- and crop-specific nutrient recommendation. Geo-statistical analysis and GIS-based soil mapping approach can largely counter the challenge through developing soil fertility maps on the wide scale by exclamation of collected data from a small number of samples. Hence, updated soil maps based on variability in distribution of native nutrients can be valuable to evaluate and precise recommend fertilizer for higher crop production and environmental protection.

As the availability of soil indigenous potash varies based on soil type and clay mineralogy, therefore, imbalanced uses of potash have diverse impact on crop production (Surapaneni et al. 2002; Srinivasarao et al. 2006). Accordingly, the blanket application of potash containing fertilizers without considering the soil K-supplying capacity can reduce crop yield and lead to K mining or wasting of resources. On a broader perspective, integrated nutrient management with spatial K recommendation is prerequisite to overcome current lower yield per unit land area of the country as compared with other neighboring Asian countries. As a matter of fact, each farmer in Pakistan cultivates one or several small field plots, and so, appropriate

A. Wakeel, M. Ishfaq, *Potash Use and Dynamics in Agriculture*,
https://doi.org/10.1007/978-981-16-6883-8_11

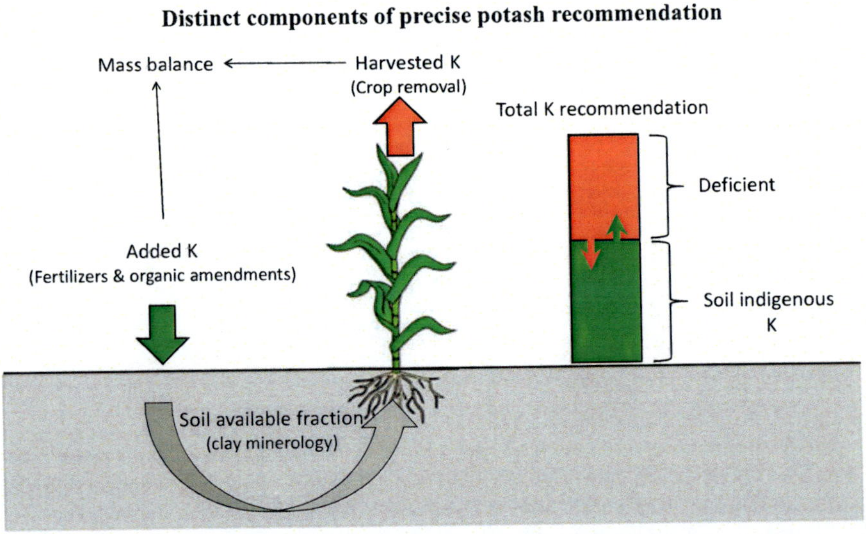

Fig. 11.1 Potash availability for plant highly depends on soil fertility status and mineralogical composition. Accordingly, K recommendation should be in balance with its net removal considering its share coming from soil indigenous reserves

nutrient management is limited owing to scare understanding of fertility status of each plot. Therefore, future strategies of nutrient management, especially for K need to be updated based on soil clay mineralogical composition and K-fixing capacity (Akhtar and Dixon 2013).

In Pakistan, it is nearly impossible to recommend fertilizer based on the soil test of each field owing to the higher small-scale farm size and insufficient soil testing facilities. Generally, the same amount of fertilizer is recommended for the region regardless of soil fertility status of each field. Accordingly, some area remains deficient, whereas other areas may receive too much fertilizer which is not economical. Therefore, K recommendation scenario can be updated consistent with site-specific manner based on unevenness in soil indigenous K reserves or K-supplying capacity of the distinct soil to sustain soil fertility and productivity (Fig. 11.1). Adequate nutrient management especially for cash or export-oriented crops based on the soil test rather than blanket application has great potential. Evidence-based decisions on K application rate, time of application, placement, and fertilizer form have resulted in the improvement in KUE, but still much remains open to elucidate.

Crop response to applied K may not be obvious in soils having larger share of available K from the non-exchangeable pool, even when these soils are rated low in exchangeable K. As exactly how much fraction of non-exchangeable K participates is still unclear, meanwhile currently no satisfactory method available in Pakistan to measure or quantify the portion of non-exchangeable K contributing to crop production (see Chap. 5). A wide variety of extractants have been used for measuring the availability of non-exchangeable K in soils such as electro-ultrafiltration (Zoerb et al.

2014). District-wise or distinct cropping system-based classification, for instance, low, medium, and high K level, in line with intensive soil sampling and GIS mapping can be valuable to recommend precise and balanced use of potash in Pakistan. Furthermore, the amount of K fertilizer to be applied should be based on specific to crops, varieties, climate, and other management practices.

The organic amendments such as animal manure, biogas residues, slurry, or food waste are also equally imperative in computing overall K balance. The potash availability may be limited due to the lack of financing, the shortage of stock, or partial demand. An integrated management effort needs to be made to meet the substantial share of K from organic manure and crop residues recycling. As crop residues recycle significant quantities of K back into the soil, removal of crop residues for other purposes, such as animal bedding or fuel without potash application, can rapidly deplete the soil K fertility. As per crop residues are source of soluble K accessible to all plant species and may, thereby, provide additional K to crops with poor access to non-exchangeable K or with only shallow root systems. It is well tested that the application of organic and inorganic fertilizers in combination can increase K use efficiency and substitute the considerable part of chemical fertilizers.

The right method of fertilizer application can be better rationalized such as foliar sprays or deep placements to get better crop response. For example, spraying of rainfed crops with KCl solution during drought or intermitted drought will be one of the drought-mitigation strategies in dryland agriculture. In addition, most K recommendations consider as a sole application; however, crop requirements of all nutrients need to be addressed to make efficient use of all inputs. The combined application of NPK has consistently proved to be advantageous in terms of increasing crop yield over just NP application. Furthermore, crop capacities to acquire K differ according to root system architecture, i.e. tap-rooted systems and fibrous systems found in crop species (Kutschera et al. 2009). Tap-rooted systems usually enable plants to access deeper horizons, while colonizing the topsoil less densely than fibrous systems.

11.2 Potential Use of MOP in Pakistan

Muriate of potash (MOP) or potassium chloride (KCl), natural white crystals mined from deep deposits, is widely used as major source of potash. Characteristic fertilizer analysis of MOP comprises of 60–63% K_2O and 46–48% chloride (Cl). The side products from KCl mining, i.e. potassium sulfate (SOP) and potassium nitrate (KNO_3), are also commercially available in Pakistan but they are comparatively expensive. These products are considered ideal for crops which are particularly sensitive to Cl, for instance, certain fruits and vegetables.

Potassium sulfate is frequently recommended as a source of K in Pakistan in consequence of its additional benefits of sulfur; nevertheless, it is nearly out of the capacity of poor smallholder Pakistani farmers due to the massive increase in its price since last few years. In Pakistani agriculture, MOP applications is verified safe and promising source of potash without residual accumulation of Cl in the root zone

(Tariq et al. 2011). Nonetheless, still comprehensive and long-term experimental data are required in the country to reveal the consequences of MOP and Cl^- application on the quality of agricultural commodities and their residual effects.

According to the local market scenario, national fertilizer companies have commenced to import MOP as alternate to SOP which is comparatively cost-effective and may be beneficial to promote potash application in Pakistan because of its lower price (nearly 2/3 of SOP), therefore, has higher potential to be adopted by local farming community. Then again, a significant knowledge gap exists to elucidate its feasibility and comparative economic return in distinct agroecosystems of Pakistan.

Long-term field experiment under local conditions can justify the feasibility of the right source of potash fertilizer with considering its use efficiency and net economic return. A considerable effect of MOP application is verified in terms of crop growth and yield. A study conducted by Khan et al. (2010) in Pakistan indicated that potato (*Solanum tuberosum* L.) tuber yield was higher when the K was applied as MOP as opposed to SOP. In another study, tomato yield increased with MOP application as compared to SOP when both sources of K were applied in the same concentration (Akhtar et al. 2010). Similarly, MOP as a source of K can be used for maize crop as it can tolerate salt (Tariq et al. 2011). Soil analysis indicate that there are no potential risks of Cl toxicity in the rhizosphere, as it has negative charge (Cl^-), thus it is very mobile in soil and can leach down from rhizosphere (Shaaban and Abou El-Nour 2012). Therefore, MOP can be used as a source of potash largely with good leaching conditions.

In addition to K, MOP is also a good source of Cl, and as Cl is an essential micronutrient beneficial for plants under various conditions, for instance, it participates in osmoregulation and disease suppression. Meanwhile, application of MOP is revealed risky in the salt-affected soils of Pakistan (Hussain et al. 2000), due to its high salt index. Additionally, chlorophobic crop species (i.e. tobacco, sugarcane, potato, tomato, strawberry, citrus, grape, cucumber, and onion) should be avoided under poor soil permeability to fertilize with MOP. However, its application under good drainage condition has no adverse effects owing to its high solubility and the movement of irrigation water in the soil.

In brief, MOP can be considered reasonably better option of K source from economic prospects. It supplies comparatively more K and its adverse effects are only noticed when it was supplied above sufficiency levels (Watanabe et al. 2017). Further, soil having high salt index and adequate precipitation, MOP can be used with application of good quality irrigation water. As MOP has been introduced in the market, therefore, long-term comprehensive comparative studies of MOP with other potash sources should be conducted to justify the potential use of MOP in arable lands of the country.

11.3 Use of Polyhalite—A Newly Introduced Multi-nutrient Natural Fertilizer

Polyhalite with chemical formula $K_2Ca_2Mg\ (SO_4){\cdot}2H_2O$ is single complex crystal fertilizer which provides four of the six plant macronutrients namely K, calcium (Ca), magnesium (Mg), sulfur (S), and trace amounts of micronutrients including zinc (Zn), boron (B), among others. It is a natural fertilizer with neutral pH and useful for many soil and crops. It is emerging as a potential organic fertilizer source because of favorable crop response, prolonged availability, and low salt index; therefore, it can be applied along crop seeds with less potential for salt damage. Furthermore, leaching of K, Ca, Mg, and S in the soil is experienced lower meanwhile the residual effect is higher for the next crop compared to the subsequent application of the equivalent sulfate salts (see Fig. 11.2).

Polyhalite as a source of K differs from many others as it is produced from a natural mineral without any chemical processing. For instance, POLY4 (Polyhalite) is a naturally occurring, low chloride compared to MOP, multi-nutrient fertilizer and certified for organic use. Research data has justified that the nutrients, for example, Ca and Mg are valuable for plant, and largely improve physical and chemical properties of the soil. Therefore, it can be categorized as an effective fertilizer for

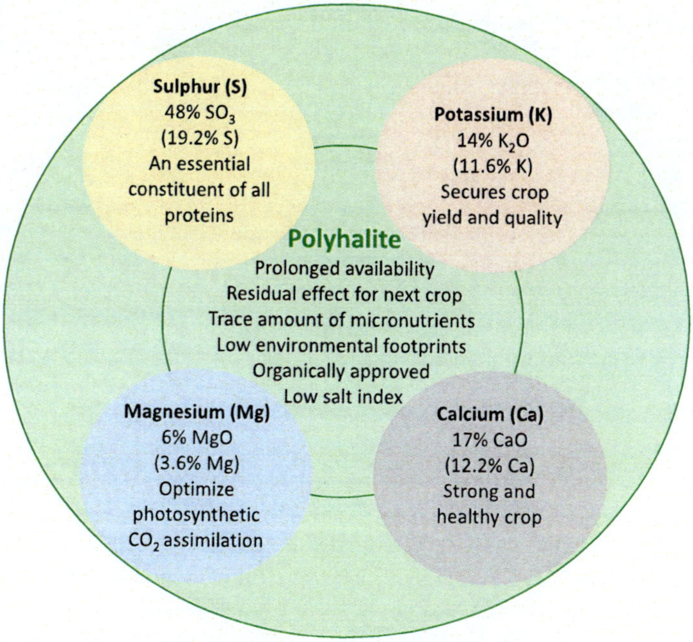

Fig. 11.2 Polyhalite—an emerging potential natural fertilizer that contains four essential plant nutrients, namely, K, Ca, Mg, S and the trace amount of valuable micronutrients

farmers to improve their crop yield, along with the quality and structure of soil with one simple product.

An important benefit of using polyhalite as a fertilizer is the rate at which it releases nutrients to the soil profile. The release of Ca, Mg, K, and S from polyhalite appears to be slower than other commonly used soluble salts (Vale 2016). It has been reported that it is less water soluble than the more conventional fertilizer sources and may conceivably provide a slower release of nutrients. However, at the normal application rate, the solubility of polyhalite is within the range of crop requirements (Yermiyahu et al. 2017). In this regard, polyhalite has the potential for long-term effects compared to other commercial fertilizers. It is revealed that POLY4 can significantly increase maize total biomass, and number of green leaves, plant height, cob height, and yield compared with MOP (Fan et al. 2017).

The application of polyhalite as a source of nutrients in the fertilization program can improve crop quality and reduce input cost. Therefore, it benefits farmers via securing their profits through improving yields with the availability of nutrient over a timeframe that tend to crop requirements, and ultimately reduce cost or both. Recently, agricultural production scenario is converting from intensive agriculture towards organic agriculture to secure sustainable food production and suppress environmental related concerns. For instance, the most recent public policy plan for sustainable agriculture development in China, including *"action to achieve zero growth of chemical fertilizer use by 2020"* and *"green food"* related emerging concepts (Xu et al. 2020). In this regard, polyhalite can be a promising source of organic fertilizer for the arable lands of Pakistan which are deficient in K, S, and Mg, and to improve NUE. However, it literally remains open to elucidate its potential compatibility under Pakistani agricultural scenario concerning following perspectives:

- Comparative economic returns with other commercially available potash sources, i.e. MOP/SOP.
- Macro and micronutrient use efficiency and legacy, and environmental footprints.
- Crop yield, quality, and disease resilience in sole or integrated (50:50) application.
- Its long-term impacts on soil health, especially in calcareous soils of the country.
- Its compatibility in saline soils or for salt sensitive fruits, vegetables, and grains.

11.4 Use of Potassium for Abiotic Stress Tolerance

The most recent climate models predict that duration and incidences of heat stress and drought are expanding. These growing manifestations of abiotic stresses, including heat stress and drought are for the most parts of the world due to decline in soil fertility in addition to global warming which negatively affecting our major crops. Being sessile in nature, plants encounter a wide range of unfavorable climatic conditions with the availability of adequate nutrients, water, and energy. In this context, potash supply revealed a promising potential to protect plants from the adverse effects of biotic and abiotic stresses (see Chap. 8).

Most research work is now focusing to manage plant nutrient in such a way to protect plant from the adverse effects of environmental stresses. In Pakistan, the significance of K nutrition to increase plant growth and yield is well addressed at various forums, nevertheless, its consequences to combat abiotic stresses and sustainable crop production is not satisfactory elucidated. It is prerequisite to comprehensively elucidate the beneficial effects of K nutrition to mitigate numerous abiotic stresses, over and above, introduce obtain results at the ground level for farmer benefits. Comparative studies for K fertilization with farmer practices may be meaningful, especially considering the significance of potash to cope with biotic and abiotic stresses which are very common in Pakistan.

The most agricultural soils in Pakistan are under arid and semi-arid regions, and the availability of irrigation water is foremost limiting factor for crop production. As prolonged plant exposure to drought stress can damage them with the generation of reactive oxygen species (ROS). In contrast, as mentioned in the previous sections that satisfactory K nutrition improves the osmotic adjustment and protects plants from the adverse effects of water stress (Egilla et al. 2005). Furthermore, due to more net evapotranspiration than average rainfall, considerable arable lands in Pakistan are salt affected. The function of K to mitigate the detrimental effects of salt stress is well reported (Cakmak 2005; Ashraf et al. 2012; Wakeel 2013; Gul et al. 2019). Additionally, an adequate K supply also enhances plant resistance to combat against low temperature stress and waterlogging (Wang et al. 2013). As consequences of K nutrition to mitigate abiotic stress is well documented, but its practical implications are still missing in farmer field of the country.

11.5 Unveiling the Effect of Potassium on Produce Quality

Research discloses that many crop quality traits are mediated by adequate potash supply with one or more other farming practices. The interactions of K with other quality effecting factors such as climatic influences, other mineral nutrients, or even crop management decisions associated with time, rate, and method are worthwhile. The rationalization of these interaction in different agroecosystems can provide systematic understanding of the consequences of K nutrition for crop quality for maximum economic yield.

One of the most prominent functions of K is to improve quality of agricultural commodities which is closely related with adequate nutrient supply (Pettigrew 2008). The synthesis of photoassimilates and their translocation from the source to sink organs are directly facilitated by K supply, and its limited supply can deteriorate the quality of grains, fruits, and vegetables. For instance, potash nutrition in citrus influences the fruit size, sweetness, peel thickness, and color (Alva et al. 2006). An adequate K supply is also experienced favorable for other crop species such as it improves protein and milling quality of wheat, oil contents in soybean, and the weight per kernel in corn (Mengel and Kirkby 2001). Furthermore, it enhances the sucrose content in sugarcane and sugar beet, quality of cotton fiber, and shelf life of fruits and vegetables (see Chap. 6). Apart from improving direct quality of agriculture produce, K reduces the incidences of diseases and insects/pests. Additionally, crop quality is also influenced indirectly as a result of lodging under K deficient conditions, and potash supply can increase stem strength and accordingly lodging resistance (Zaman et al. 2015). In addition, plants resist against wilting and frost also improves with K fertilization, which ultimately maintains crop yield and quality (see Chap. 8).

In Pakistan, very little attempts have been made to highlight the quality traits of agricultural commodities. There is high competition in the global market, thus quality of agriculture produce is imperative to be consider for export to uplift country economy. The export of agricultural commodities and their quality-based price demands balanced fertilization, expressly adequate K nutrition. For example, the use of potash improves the taste, color, flavor, and shelf life of fruits which have more economic value. Therefore, to promote the consequences of K nutrition to improve the crop quality, standards for comparison and execution are required in Pakistan. These standards differ with the crop, customer preference, manufacturing specifications, the intended use of agriculture commodities and their nutritional value. Systematic management practices, satisfactory economic incentives, and favorable environmental conditions can make possible to utilize maximum local potential.

11.6 Potash to Improve Nitrogen Use Efficiency by Plants

Combined application of fertilizers with potash is an efficient and cost-effective approach to attain higher N use efficiency (NUE). Sufficient K supply improves both N uptake by plant roots and its assimilation to amino acids and protein, consequently increase crop yield. According to International Potash Institute (IPI) findings, balanced K nutrition can instantly improve NUE nearly 20% in different crop species. In several intensive farming areas, N is overused with intentions to accomplish maximum crop yields which usually result in low NUE and ultimately undesirable environmental concerns (Zhang et al. 2015; Shahzad et al. 2019). Meanwhile, potash fertilization in proper ratio with N improves crop yields and NUE, which bring about less residual nitrate (NO_3^-) in soils and consequently reduce its susceptibility to contaminate ground water (Fig. 11.3).

In accordance with numerous-field trials collaborated by IPI, it is revealed that at low soil K fertility levels, an increase in N supply depressed yields of barley, rice, and cassava. Therefore, potash application is essential for better utilization of N and P and to maintain soil K fertility (White and Brown 2010; Timilsena et al. 2015). For example, the split applications of K in wheat can increase N uptake and accordingly grain yields (Lu et al. 2014). In addition, K facilitates assimilation of NO_3^- in the plant aerial parts via enhancing the activities of NO_3^- assimilating enzymes resulting in improves in N assimilation in the plant and synthesis of protein (Fig. 11.2). For instance, K applications increased the quality of pearl millet by increasing protein

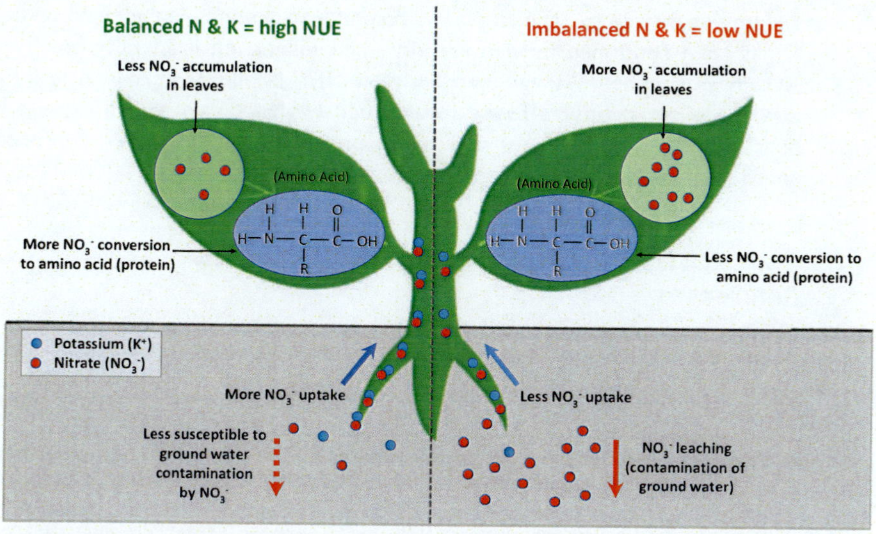

Fig. 11.3 A simplified illustration of the uptake, translocation, and assimilation of nitrate in plant mediated by K availability. Adequate K supply improves both NO_3^- absorption and its conversion to amino acids and protein in plants, accordingly, improves quantity and quality of food production, meanwhile reduces its environmental footprints

content and yield, indicating better utilization of applied N. The maximum physiological attributes, carbohydrate and the juice extract of sweet sorghum is reported with the combined application of N and K (Almodares et al. 2008).

Plant uptake and translocate K as a counter-ion together with NO_3^- in xylem to the aerial parts. Plants can possibly sense the level of N and K in the soil solution and regulate correspondingly to take up and transport them from root-to-shoot to regulate their distribution between various organs (Raddatz et al. 2020). In plant, NO_3^- first reduced to amines and then combined into amino acids to finally form proteins. Meanwhile, restricted K supply suppress the optimum NO_3^- transport and slow down protein synthesis, followed by the accumulation of nitrate-N and soluble amino-N in the plant. Therefore, K nutrition facilitates to maintain NO_3^- content in agricultural commodities within certain limits proposed by the public health authorities. So adequate K supply, on the one hand, reduces NO_3^- accumulation in plant tissue with optimum protein synthesis which conforms the consumer's concern for "safe food," on the other hand, the higher NO_3^- uptake fulfills the prerequisite for environmentally-friendly crop production.

In Pakistan, the consumption of K fertilizer for agricultural purpose is much lower (nearly 1%) than that of N which lead to soil K depletion and low NUE (see Chap. 9 for detail description). Higher applications of sole N or K do not certainly make possible to improve economic return and may even decrease yield. The interaction of N-K has potential to cope with current concern of low NUE and its concomitant threats, therefore, setting optimal N: K nutritional ratios can be profitable for high yielding quality crops. The crop yield response with combined application of N and K has been confirmed in many reports, and interaction is literally positive when NO_3–N is supplied. Antagonism between NH_4–N and K in uptake is frequently justified as competitive effects; even though evidence exist which express mixed-noncompetitive interactions. However, precise research studies are needed to uncover underlying mechanisms.

References

Akhtar ME, Khan MZ, Rashid MT, Ahsan Z, Ahmad S (2010) Effect of potash application on yield and quality of tomato (*Lycopersicon esculentum* mill.). Pak J Bot 42(3):1695–1702

Akhtar MS, Dixon JB (2013) Mineralogical characteristics and potassium quantity/intensity relation in three Indus river basin soils. Asian J Chem 21:3427–3442

Almodares A, Taheri R, Chung IM, Fathi M (2008) The effect of nitrogen and potassium fertilizers on growth parameters and carbohydrate contents of sweet sorghum cultivars. J Environ Biol 29: 849–852

Alva AK, Mattos JD, Paramasivam S, Patil B, Dou H, Sajwan K (2006) Potassium management for optimizing citrus production and quality. Int J Fruit Sci 6:3–43

Ashraf M, Afzal M, Ahmad R, Maqsood MA, Shahzad SM, Tahir MA, Akhtar N, Aziz A (2012) Growth response of the salt-sensitive and the salt-tolerant sugarcane genotypes to potassium nutrition under salt stress. Arch Agronomy Soil Sci 58:385–398

Cakmak I (2005) The role of potassium in alleviating detrimental effects of abiotic stresses in plants. J Plant Nutr Soil Sci 168:521–530

Egilla JN, Davies FT, Boutton TW (2005) Drought stress influences leaf water content, photosynthesis, and water-use efficiency of hibiscus rosa-sinensis at three potassium concentrations. Photosynthetica 43:135–140

Fan, X., Meakin, R., Pavuluri, K., Lewis, T. D. and Qin, Y. 2017. Effect of polyhalite and MOP as a K source on the yield and nutrient uptake of maize in Sichuan, China. ASA, CSSA, SSSA 2017 Annual Meeting, 22–25 October 2017. Abstract 340-10. Available at: https://scisoc.confex.com/crops/2017am/webprogram/Paper10874 0.html

Gul M, Wakeel A, Steffens D, Lindberg S (2019) Potassium-induced decrease in cytosolic Na^+ alleviates deleterious effects of salt stress on wheat (*Triticum aestivum* L.). Plant Biol 21(5):825–831

Hussain T, Ashraf M, Abbas MA, Ali TI (2000) Comparative effectiveness of two potassium sources in rice-wheat cropping system. Pak J Agric 16(1):17–19

Khan MZ, Akthar ME, Safdar MN, Mahmood MM, Ahmad S, Ahmad N (2010) Effect of source and level of potash on yield and quality of potato tubers. Pak J Bot 42(5):3137–3145

Kutschera L, Lichtenegger E, Sobotik M (2009) Wurzelatlas der Kulturpflanzen gemäßigter Gebiete mit Arten des Feldgemüsebaues. DLG Verlag, Frankfurt/Main

Lu Q, Jia D, Zhang Y, Dai X, He M (2014) Split application of potassium improves yield and end-use quality of winter wheat. Agron J 106(4):1411–1419

Mengel K, Kirkby EA (2001) Principles of plant nutrition. Dordrecht, Boston, London, Kluwer Academic Publishers

Pettigrew WT (2008) Potassium influences on yield and quality production for maize, wheat, soybean and cotton. Physiol Plant 133:670–668

Raddatz N, de los Ríos LM, Lindahl M, Quintero FJ, Pardo JM (2020) Coordinated transport of nitrate, potassium, and sodium. Front Plant Sci 11:247

Shaaban SHA, Abou El-Nour EAA (2012) Effect of different potassium sources on yield and nutrient uptake by flax (*Linum usitatissimum* L.) grown on loamy sand soil. J Appl Sci Res 8 (3):1425–1429

Shahzad AN, Qureshi MK, Wakeel A, Misselbrook T (2019) Crop production in Pakistan and low nitrogen use efficiencies. Nat Sustain 2(12):1106–1114

Srinivasarao C, Rupab TR, Subba RA, Ramesha G, Bansald SK (2006) Release kinetics of non-exchangeable potassium by different ex-tractants from soils of varying mineralogy and depth. Soil Sci Plant Anal 37(3):473–449

Surapaneni A, Palmer AS, Tillman RW, Kirkman JH, Gregg PEH (2002) The mineralogy and potassium supplying power of some loessial and related soils of New Zealand. Geoderma 110: 191–204

Tariq M, Saeed A, Nasir M, Mian IA (2011) Effect of potassium rates and sources on the growth performance and on chloride accumulation of maize in two different textured soils of Haripur, Hazara division. Sarhad J Agric 27(3):415–422

Timilsena YP, Adhikari R, Casey P, Muster T, Gilla H, Adhikari B (2015) Enhanced efficiency fertilisers: a review of formulation and nutrient release patterns. J Sci Food Agric 95(6):1131–1142

Vale, F. 2016. Calcium and magnesium movement in soil profile with polyhalite as potassium fertilizer for soybean crop. Proceedings of FERTBIO 2016, Goiana, Brazil, October, 16–20.

Wakeel A (2013) Potassium–sodium interactions in soil and plant under saline-sodic conditions. J Plant Nutr Soil Sci 176(3):344–354

Wang M, Zheng Q, Shen Q, Guo S (2013) The critical role of potassium in plant stress response. Int J Mol Sci 14:7370–7390

Watanabe K, Tominaga J, Yabuta S, Takaragawa H, Suwa R, Ueno M, Kawamitsu Y (2017) Effects of different kinds of potassium and chloride salts on sugarcane quality and photosynthesis. Sugar Technol 19(4):8–385

White PJ, Brown PH (2010) Plant nutrition for sustainable development and global health. Ann Bot 105(7):1073–1080

Xu J, Zhang Z, Zhang X, Ishfaq M, Zhong J, Li W, Zhang F, Li X (2020) Green food development in China: experiences and challenges. Agriculture 10(12):614

Yermiyahu U, Zipori I, Faingold I, Yusopov L, Faust N, Bar-Tal A (2017) Polyhalite as a multi nutrient fertilizer - potassium, magnesium, calcium and sulfate. Israel J Plant Sci 64 (3–4):145–157

Zaman U, Ahmad Z, Farooq M, Saeed S, Ahmad and Wakeel, A. (2015) Potassium fertilization may improve stem strength and yield of basmati rice grown on nitrogen-fertilized soils. Pak J Agric Sci 52:439–445

Zhang X, Davidson EA, Mauzerall DL, Searchinger TD, Dumas P, Shen Y (2015) Managing nitrogen for sustainable development. Nature 528:51–59

Zoerb C, Senbayram M, Peiter E (2014) Potassium in agriculture-status and perspectives. J Plant Physiol 171:656–669

Index

A. Wakeel, M. Ishfaq, *Potash Use and Dynamics in Agriculture*,
https://doi.org/10.1007/978-981-16-6883-8